From Static to Dynamic Couplings in Consensus and Synchronization among Identical and Non-Identical Systems

Von der Fakultät Konstruktions-, Produktions-, und Fahrzeugtechnik
der Universität Stuttgart zur Erlangung der Würde eines
Doktors der Ingenieurwissenschaften (Dr.–Ing.) genehmigte Abhandlung

Vorgelegt von

Peter Wieland

aus Stuttgart–Bad Cannstatt

Hauptberichter: Prof. Dr.–Ing. Frank Allgöwer
Mitberichter: Prof. Dr. Alberto Isidori
Prof. Dr. Rodolphe Sepulchre

Tag der mündlichen Prüfung: 06.09.2010

Institut für Systemtheorie und Regelungstechnik
Universität Stuttgart

2010

Bibliografische Information der Deutschen Nationalbibliothek

Die Deutsche Nationalbibliothek verzeichnet diese Publikation in der Deutschen Nationalbibliografie; detaillierte bibliografische Daten sind im Internet über http://dnb.d-nb.de abrufbar.

D 93

ISBN 978-3-8325-2638-2

Logos Verlag Berlin GmbH
Comeniushof, Gubener Str. 47,
10243 Berlin
Tel.: +49 (0)30 42 85 10 90
Fax: +49 (0)30 42 85 10 92
INTERNET: http://www.logos-verlag.de

Für Melanie, Lea und Anna

Acknowledgments

The results presented in this thesis are the outcome of the research I performed during my time as a research assistant at the Institute for Systems Theory and Automatic Control (IST), University of Stuttgart in the years 2005 to 2010. Throughout this exciting journey I had the chance to meet and interact with many interesting and inspiring people. The present thesis was positively influenced by many of them and I wish to convey my sincere gratefulness to them.

First, I want to express my gratitude towards my supervisor, Prof. Frank Allgöwer. He endowed me with lots of freedom in my research and offered me the unique opportunity to be part of a very open-minded, internationally renowned research group. I also want to thank Prof. Rodolphe Sepulchre, with whom I had the chance to have a very fruitful interaction during the last year of my thesis, for sharing his knowledge and insight with me in a very motivating way.

Moreover, I am deeply indebted to Prof. Alberto Isidori, Prof. Rodolphe Sepulchre, and Prof. Peter Eberhard for their interest in my research and for accepting to be members of my doctoral exam committee. In addition, I thank Prof. Christian Ebenbauer for accepting to stand in for Prof. Isidori, who could unfortunately not be present at the oral exam.

My time as a research assistant became a rich experience by interacting with my colleagues at the IST. I shared the interest in research on interconnected dynamical systems with Rainer Blind, Jørgen K. Johnsen, Prof. Jung-Su Kim, Ulrich Münz, and Gerd S. Schmidt. They contributed with many vivid discussions and useful hints to the results presented in this thesis. Furthermore, I want to thank Tobias Raff, with whom I shared office most of the time, for letting me have part in his experience as a researcher and taking care of me as a novice.

In addition to those already mentioned, I wish to thank my colleagues Christoph Böhm, Christian Breindl, Jan Hasenauer, Christoph Maier, Marcus Reble, and Simone Schuler as well as Hongkeun Kim from Seoul National University, Korea and Tilman Utz from TU Wien, Austria for their helpful comments on this thesis.

Besides those mentioned explicitly, I am indebted to all of my current and former colleagues at the IST for their generous support, uncounted discussions, and lots of fun in academic as well as non-academic interactions. It was a great pleasure being part of this group.

Last but not least, I want to thank my parents for their support in all these years, I want to thank Melanie for her love, patience, and encouragements, and I want to apologize to Lea for all the time she was not allowed to disturb me and I was unavailable for playing.

Stuttgart, September 2010
Peter Wieland

Contents

Symbols and Acronyms

Symbols

$\mathbb{R}$	set of real numbers
$\mathbb{R}_+$	set of positive real numbers
$\overline{\mathbb{R}}_+$	set of non-negative real numbers
$\overline{\mathbb{R}}_-$	set of non-positive real numbers
$\mathbb{C}$	set of complex numbers
j	the imaginary unit ($\mathrm{j} = \sqrt{-1}$)
$\mathrm{j}\mathbb{R}$	set of imaginary numbers
$\overline{\mathbb{C}}_+$	the closed right-half complex plane
$\mathbb{C}_-$	the open left-half complex plane
$\overline{\mathbb{C}}_-$	the closed left-half complex plane
$\mathbb{N}$	the natural numbers
$\mathbb{N}_k$	the natural numbers from 1 to k ($\mathbb{N}_k = \{1, \dots, k\} \subset \mathbb{N}$)
$\mathbb{Z}$	the integers
$\mathbb{S}^1$	the circle
$\mathfrak{Re}(z)$	the real part of a complex number $z \in \mathbb{C}$
$\mathfrak{Im}(z)$	the imaginary part of a complex number $z \in \mathbb{C}$
$\overline{z}$	the complex conjugate of $z \in \mathbb{C}$
I_n	the $n \times n$ identity matrix
1_n	the n-dimensional vector of ones
$\otimes$	the Kronecker product of two matrices (cf. Horn and Johnson, 1991)
$\sigma(A)$	the spectrum (i.e., set of eigenvalues) of a square matrix $A \in \mathbb{R}^{n\times n}$ or $A \in \mathbb{C}^{n\times n}$
$\lambda_k(A)$	the kth eigenvalue of a square matrix $A \in \mathbb{R}^{n\times n}$ or $A \in \mathbb{C}^{n\times n}$ with the convention that $\mathfrak{Re}(\lambda_k(A)) \leq \mathfrak{Re}(\lambda_{k+1}(A))$, $k \in \mathbb{N}_{n-1}$
$\rho(A)$	the spectral radius of a square matrix $A \in \mathbb{R}^{n\times n}$ or $A \in \mathbb{C}^{n\times n}$
A^H	the complex conjugate transpose of some matrix $A \in \mathbb{C}^{n\times m}$
A^+	the Moore-Penrose generalized inverse of some matrix $A \in \mathbb{R}^{n\times m}$ or $\mathbb{C}^{n\times m}$ (cf. Horn and Johnson, 1985); if $\operatorname{rank}(A) = n$, $A^+ = A^T(AA^T)^{-1}$ is a right-inverse of A; if $\operatorname{rank}(A) = m$, $A^+ = (A^TA)^{-1}A^T$ is a left-inverse of A
$\ker(A)$	the kernel of a matrix or linear map A
$\operatorname{im}(A)$	the image of a matrix or linear map A
$\operatorname{span}(v_1, \dots, v_N)$	the subspace of $\mathbb{R}^n$ spanned by the vectors $v_k \in \mathbb{R}^n$, $k \in \mathbb{N}_N$

$\mathrm{diag}(x)$	for a vector $x \in \mathbb{R}^n$ the $n \times n$ diagonal matrix with the elements of x on the diagonal
$\mathcal{X}/\mathcal{Y}$	the factor space of equivalence classes $x + \mathcal{Y}$, $x \in \mathcal{X}$ for a linear vector space $\mathcal{X}$ and a subspace $\mathcal{Y} \subset \mathcal{X}$ (see Appendix C)
$A/\mathcal{Y}$	the map induced by the linear map $A : \mathcal{X} \to \mathcal{X}$ on the factor space $\mathcal{X}/\mathcal{Y}$ with $A\mathcal{Y} \subset \mathcal{Y}$, defined as $x + \mathcal{Y} \mapsto (Ax) + \mathcal{Y}$ (see Appendix C)
$A\lvert\mathcal{Y}$	the linear map $A : \mathcal{X}_1 \to \mathcal{X}_2$ restricted to the subspace $\mathcal{Y} \subset \mathcal{X}_1$; the map $A\lvert\mathcal{Y} : \mathcal{Y} \to A\mathcal{Y}$ is defined as $x \mapsto Ax$, $x \in \mathcal{Y}$
$\lvert\mathcal{V}\rvert$	the cardinality of a discrete set $\mathcal{V}$
$\lvert x\rvert$	the magnitude or absolute value of $x \in \mathbb{R}$ or $x \in \mathbb{C}$
$\lVert x\rVert$	norm of some vector $x \in \mathbb{R}^n$ or $x \in \mathbb{C}^n$; unless otherwise stated, $\lVert x\rVert$ indicates the 2-norm or Euclidean norm of x
$\lVert A\rVert$	norm of some matrix $A \in \mathbb{R}^{m\times n}$ or $A \in \mathbb{C}^{m\times n}$; unless otherwise stated, $\lVert A\rVert$ indicates the operator norm of A induced by the 2-norm
$\mathrm{conv}(S)$	the convex hull of some set S, e.g., $S \subset \mathbb{C}$
$\succ, \succeq, \prec, \preceq$	greater, greater equal, less, and less equal with respect to the partial ordering induced by the positive semidefinite cone in $\mathbb{R}^{n\times n}$ or $\mathbb{C}^{n\times n}$ in matrix inequalities
$\omega(x)$, $\omega(\mathcal{B})$	ω-limit set of a point $x \in \mathbb{R}^n$ or a set $\mathcal{B} \subset \mathbb{R}^n$ (see Appendix B)
$L_f h(x)$	the directional derivative of $h : \mathbb{R}^n \to \mathbb{R}^m$ along a vector field $f : \mathbb{R}^n \to \mathbb{R}^n$: $L_f h(x) \triangleq \frac{\partial h(x)}{\partial x} f(x)$ (see Isidori, 1995)
$\triangleq$	definition; $A \triangleq B$ defines A to be equal to B

Acronyms

FBI	Francis-Byrnes-Isidori
iPRC	infinitesimal phase response curve
iSCC	independent strongly connected component
LMI	linear matrix inequality
LQR	linear quadratic regulator
LTI	linear time-invariant
ODE	ordinary differential equation

Abstract

Consensus and synchronization problems have been popular subjects in systems and control theory over the last couple of years, mainly motivated by the fact that phenomena summarized under these two terms are observed in various instances in a wide range of scientific disciplines. The two terms both refer to the property that individuals in a group reach agreement in some sense. In addition, couplings are typically of a diffusive type in both cases, i.e., the individual systems exchange only relative information.

We will consider consensus and synchronization in networks of individual dynamical systems interconnected according to a specific communication topology, where the individual systems are modeled by ordinary differential equations and the communication topology is modeled by a graph. Typically, consensus problems deal with simple individual system dynamics and weak assumptions on the communication graph. In contrast, synchronization problems commonly focus on complex individual system dynamics and simple communication topologies. In both cases, couplings are typically static and of a diffusive type. Yet, very few results exist that consider complex individual systems and complex communication topologies at the same time. There seems to be a tradeoff between admissible system complexity and admissible topological complexity in consensus and synchronization problems.

In view of this observation, we address two questions in this thesis. Firstly, we ask for the reasons for this tradeoff, i.e., we ask for the limitations that are inherent to static diffusive couplings. Secondly, we ask to what extent these limitations can be removed by appropriately extending static diffusive couplings.

We show that weak assumptions on the communication graph yield strong requirements imposed on the acting and sensing capabilities of the individual systems as well as their stability properties. On top of this tradeoff between system and topological complexity, static diffusive couplings are generally not suited to achieve consensus or synchronization if the individual systems admit non-identical dynamical models.

We propose diffusive couplings that are extended by dynamic compensators to overcome some of the aforementioned limitations of static diffusive couplings.

Strong requirements on the sensing capabilities are partly removed with the help of dynamic observers. The observer design problem for consensus problems is addressed subject to a relative sensing constraint in this thesis. We explain that this constraint yields a state estimation problem with unknown inputs and propose a solution based on unknown-input observers.

In case of heterogeneous networks, we show that the satisfaction of an internal model principle is necessary for consensus and synchronization. We argue that this principle can generally only be satisfied if dynamic couplings are employed. The internal model principle derived in this thesis is related and compared to the theory of output regulation with its well-known internal model principle of control theory. Thereby we establish a link between consensus and synchronization on the one hand and the theory of output regulation on the other hand.

Eventually, we are able to give solutions to consensus and synchronization problems for arbitrary heterogeneous linear networks and for networks of heterogeneous nonlinear oscillators. We thereby remove most of the limitations mentioned above and thus allow for consensus and synchronization in networks with increased system and topological complexity.

Deutsche Kurzfassung

Motivation und grundsätzliche Fragestellungen

Konsens und Synchronisation sind weit verbreitete Phänomene, die in zahlreichen verschiedenen Wissenschaftsdisziplinen eine Rolle spielen. Konsens- und Synchronisationsprobleme treten z.B. in den Sozialwissenschaften, der Biologie, der Physik sowie den Ingenieurwissenschaften auf (Pikovsky et al., 2001, Strogatz, 2003). Vor diesem Hintergrund trägt die vorliegende Arbeit zu einem grundlegenden systemtheoretischen Verständnis einiger Mechanismen bei, die zu Konsens und Synchronisation führen.

Sowohl Konsens als auch Synchronisation bezeichnen in der Systemtheorie ein spezifisches strukturiertes Verhalten in einer Gruppe von dynamischen Systemen. Beide Begriffe beschreiben die Eigenschaft, dass die einzelnen Systeme in der Gruppe eine „Einigung" über bestimmte Zustands- oder Ausgangsgrößen erreichen. Die betrachtete Größe in Konsensproblemen ist in der Regel der konkrete Wert, auf den sich die einzelnen Systeme einigen, während sich Synchronisationsprobleme eher auf die Zeitpunkte beziehen, zu denen bestimmte Werte erreicht werden. Diese Unterscheidung ist jedoch in erster Linie eine Frage des Standpunktes, da Konsens bezüglich Zeitpunkten zu Synchronisation und Synchronisation von Systemtrajektorien zu Konsens führt. Konsens und Synchronisation beschreiben also im Wesentlichen dasselbe strukturierte Verhalten in Gruppen dynamischer Systeme. Dennoch werden die Begriffe in der Regel sehr unterschiedlich verwendet, wie im Folgenden dargelegt wird.

Eine wesentliche Voraussetzung, um Konsens oder Synchronisation in einem Netzwerk aus einzelnen Systemen zu erreichen, ist ein bestimmtes Maß an Kopplungen zwischen den Teilsystemen des Netzwerks. Häufig entstehen diese Kopplungen durch einen gewissen Informationsaustausch, d.h. im weitesten Sinne durch Kommunikation zwischen den Teilsystemen. Diese Kommunikation erfolgt zumeist gemäß einer spezifischen Kommunikationstopologie. In diesem Fall können drei wesentliche Quellen von Komplexität im Gesamtnetzwerk unterschieden werden: Systemkomplexität, topologische Komplexität und Kopplungskomplexität. Die Systemkomplexität wird durch die Komplexität der Dynamik der einzelnen Teilsysteme festgelegt, die topologische Komplexität beschreibt welche Systeme miteinander kommunizieren können und die Kopplungskomplexität berücksichtigt, dass die Kommunkationskanäle zwischen den einzelnen Systemen nicht ideal sind, sondern z.B. Übertragungstotzeiten existieren oder Information in Form von Paketen übertragen wird. In der vorliegenden Arbeit werden nur die beiden ersten Komplexitätsdimensionen berücksichtigt (Scardovi and Sepulchre, 2009, siehe auch), welche insbesondere erlauben, eine Unterscheidung zwischen klassischen Konsensproblemen einerseits und klassischen Synchronisationsproblemen andererseits wie folgt vorzunehmen:

Konsensprobleme beschäftigen sich in der Regel mit hoher topologischer Komplexität, während die einzelnen Systeme durch sehr einfachen dynamischen Modelle beschrieben werden. Die hohe topologische Komplexität wird dabei durch schwache einschränkende Annahmen an die Kommunikationstopologie widergespiegelt (Moreau, 2005). Im Gegensatz dazu beschäftigt sich Synchronisation mit komplexen Teilsystemen, z.B. nichtlinearen Oszillatoren (Stan and Sepulchre, 2007, Wang and Slotine, 2005), während die topologische Komplexität häufig sehr gering ist. Oft wird der extreme Fall betrachtet, dass es gar keine Einschränkungen an die Kommunikationstopologie gibt.

Seither existieren nur wenige Ergebnisse für Netzwerke mit hoher topologischer Komplexität, wie Sie typischerweise in Konsensproblemen betrachtet wird, und gleichzeitig hoher Systemkomplexität, die häufig Gegenstand von Synchronisationsproblemen ist. In dieser Arbeit wird daher zum einen untersucht, inwieweit sich hohe topologische Komplexität und hohe Systemkomplexität unter Verwendung bestimmter Kopplungsmechanismen gegenseitig ausschließen, und zum anderen wird der Frage nachgegangen, welche Erweiterungen bestehender Kopplungsmechanismen geeignet sind, um Konsens und Synchronisation in Systemen mit größerer Systemkomplexität und gleichzeitig größerer topologischer Komplexität zu ermöglichen.

Einführung in die Lösungsansätze der Arbeit

Typischerweise werden sowohl in Konsens- als auch in Synchronisationsproblemen statische diffuse Kopplungen zwischen den einzelnen Teilsystemen betrachtet (Hale, 1997). Beispiele hierfür sind das Vicsek Modell (Vicsek et al., 1995), das häufig als Prototyp für Konsensprobleme angesehen wird, sowie das Kuramoto Modell (Kuramoto, 1975), welches ein generisches Modell für Synchronisation von Oszillatoren darstellt. Diffuse Kopplungen zeichnen sich dadurch aus, dass sie nur von relativen Größen zwischen den einzelnen Teilsystemen abhängen. Wenn die relevante Größe z.B. eine Position darstellt, werden diffuse Kopplungen nur mit Hilfe relativer Abstände realisiert und sind daher unabhängig von absoluten Positionen. Dies impliziert insbesondere, dass diffuse Kopplungen verschwinden, sobald Konsens bzw. Synchronisation zwischen den einzelnen Systemen im Netzwerk erreicht ist. Diffuse Kopplungen werden als statisch bezeichnet, wenn sie nur vom aktuellen Wert der relativen Größen abhängen und es keinen eigenen Kopplungszustand gibt. In diesem Fall wird der Zusammenhang zwischen den Kopplungen und den jeweiligen relativen Systemgrößen durch eine algebraische Gleichung beschrieben.

Zur Modellierung der Kommunikationstopologie sind Graphen ein geeignetes Mittel (Godsil and Royle, 2004). Dies gilt insbesondere im Zusammenhang mit diffusen Kopplungen. Graphen, welche die Kommunikationstopologie eines Netzwerkes beschreiben, werden als Kommunikationsgraphen bezeichnet. Insbesondere die algebraische Graphentheorie stellt zahlreiche wichtige Methoden und Werkzeuge zur Modellierung und Analyse der Kommunikationstopologie in Konsens- und Synchronisationsproblemen zur Verfügung. Diese Werkzeuge erlauben es zum einen, Aussagen über die gegenseitige Einflussnahme einzelner Systeme im Netzwerk zu machen und dadurch vorherzusagen, wie sich das synchronisierte Netzwerk verhalten wird, und zum anderen ermöglichen Sie die topologische Komplexität des Netzwerkes durch Annahmen an einen Graphen und dessen Eigenschaften auszudrücken. Dabei spielen vor allem die Eigenschaften „stark zusammenhängend“ (strongly connected), „zusammenhängend“ (connected) und „gleichmäßig zusammenhängend“ (uniformly connected) eine wichtige Rolle. Dabei ist „stark zusammenhängend" die stärkste Eigenschaft und „gleichmäßig zusammenhängend" die schwächste Eigenschaft.

Mit Hilfe der Kommunikationsgraphen sind wir in der Lage, Konsens in Netzwerken identischer linearer zeitinvarianter Systeme mit statischen diffusen Kopplungen zu untersuchen. Es stellt sich heraus, dass schwache Annahmen an den Kommunikationsgraphen – hier sind insbesondere gleichmäßig zusammenhängende Graphen zu erwähnen – zu erheblichen Einschränkungen an die dynamischen Modelle der einzelnen Systeme führen. Diese Einschränkungen betreffen sowohl Stabilitätseigenschaften als auch benötigte Sensoren und Aktuatoren. Um Konsens über statische diffuse Kopplungen und gleichmäßig zusammenhängende Graphen zu erreichen, müssen die Eingangs- und Ausgangsdimensionen der einzelnen Systeme der Zustandsdimension entsprechen und die Systemdynamik darf keine exponentiell instabilen Anteile aufweisen (Scardovi and Sepulchre, 2009). Diese starken Annahmen können zum Teil abgeschwächt werden, indem zusammenhängende oder stark zusammenhängende Kommunikationsgraphen angenommen werden. Es ist jedoch nicht möglich, die Einschränkungen an die einzelnen Teilsysteme und gleichzeitig die Annahmen an den Kommunikationsgraphen gering zu halten.

Um diese Begrenzung der Gesamtkomplexität des Netzwerks aufzuheben ist es in der Regel nötig, die statischen diffusen Kopplungen durch lokale dynamische Regler zu ergänzen, was zu dynamischen diffusen Kopplungen führt. Eine naheliegende Erweiterung besteht im Einsatz dynamischer Beobachter zur Zustandsschätzung. Im Zusammenhang mit diffusen Kopplungen müssen allerdings relative Zustände geschätzt werden. Dies führt zu zusätzlichen Herausforderungen, die in der vorliegendend Arbeit mit Hilfe von Beobachtern mit unbekannten Eingängen gelöst werden (Darouach et al., 1994, Willems and Commault, 1981). Dabei kommen Beobachter mit voller und mit reduzierter Ordnung zum Einsatz.

Ein weiterer wesentlicher Beitrag der vorliegenden Arbeit besteht in der Berücksichtigung von heterogenen Netzwerken, d.h. Netzwerke deren einzelne Teilsysteme nicht-identische Dynamiken aufweisen. Zunächst wird das Innere-Modell-Prinzip der Synchronisation als notwendige Bedingung für Synchronisation nicht-identischer linearer und nichtlinearer Systeme hergeleitet. Es wird gezeigt, dass dieser Bedingung im Allgemeinen nur genügt werden kann, wenn die Kopplungen durch dynamische Anteile ergänzt werden. Die Lösungsansätze werden mit dem klassischen Inneren-Modell-Prinzip der Regelungstechnik verglichen und in diesem Zusammenhang interpretiert. Dadurch wird eine enge Verbindung zwischen Konsens- und Synchronisationsproblemen einerseits und Problemen der Ausgangsfolgeregelung andererseits hergestellt.

Das Innere-Modell-Prinzip der Synchronisation führt schließlich auf einen neuen hierarchischen Lösungsansatz des Synchronisationsproblems in heterogenen Netzwerken. Es werden konkrete Lösungen für zwei Fälle vorgestellt. Diese sind zum einen Netzwerke, die aus beliebigen stabilisierbaren und entdeckbaren linearen zeitinvarianten Systemen bestehen, und zum anderen Netzwerke, die aus nichtlinearen exponentiell stabilen Oszillatoren aufgebaut sind. Es stellt sich heraus, dass die dynamischen Kopplungen, die in den beiden Fällen zum Einsatz kommen, nicht nur nötig sind um die unterschiedlichen Dynamiken der einzelnen Systeme zu kompensieren, sondern zusätzlich auch geeignet sind, um die oben genannte Begrenzung der Gesamtkomplexität in homogenen Netzwerken teilweise aufzuheben.

Forschungsbeiträge und Gliederung der Arbeit

Die nachfolgende Übersicht erläutert die Gliederung der Dissertation und führt die wichtigsten Forschungsbeiträge der einzelnen Kapitel auf.

Kapitel 2 – Graph Theory for Consensus Problems (Graphentheorie für Konsensprobleme)

In diesem Kapitel werden die grundlegenden Definitionen und Eigenschaften von Graphen dargelegt, sofern sie für Konsens- und Synchronisationsprobleme relevant sind (Wieland et al., 2008, 2010a). Insbesondere werden die Eigenschafte *zusammenhängend*, *stark zusammenhängend* und *gleichmäßig zusammenhängend* diskutiert. Darüber hinaus werden Interpretationen erörtert, die es erlauben, einen tiefgreifenden Zusammenhang zwischen Graphentheorie einerseits und Konsens- und Synchronisationsproblemen andererseits herzustellen. Besonderes Augenmerk liegt in diesem Kapitel auf der Laplace-Matrix eines Graphen, und insbesondere deren Eigenstruktur, deren Interpretation schließlich zur Erweiterung von Komponenten eines ungerichteten Graphen zu unabhängigen stark zusammenhängenden Komponenten von gerichteten Graphen führt.

Kapitel 3 – Consensus with Static Diffusive Couplings (Konsens mit statischen diffusen Kopplungen)

Dieses Kapitel behandelt zum einen bereits existierende Ergebnisse für Konsens mit statischen diffusen Kopplungen und enthält zum anderen zahlreiche Erweiterungen und Verbesserungen (Wieland et al., 2008, 2010a). Die einzelnen Systeme sind in diesem Kapitel durch identische lineare zeitinvariante Differenzialgleichungssysteme gegeben. Insbesondere werden die Wechselwirkungen zwischen Systemeigenschaften und Eigenschaften der Kommunikationstopologie herausgestellt, und es wird ein auf linearen Matrizenungleichungen basierender Entwurf der Verstärkungsfaktoren in den Kopplungen vorgeschlagen. Neben Konsens für konstante und zeitvariante Kopplungen in homogenen Netzwerken wird auch das Konsens-Problem mit Anführer erörtert, wobei die Dynamik des Anführers von der Dynamik der restlichen Systeme abweichen darf. Das Ergebnis schließt insbesondere auch eine Methode ein, um einen Anführer in ein bestehendes Netzwerk zu integrieren.

Kapitel 4 – Observer-Based Output Consensus with Relative Output Sensing (Beobachterbasierter Ausgangskonsens mit relativen Ausgangsmessungen)

Dieses Kapitel diskutiert den Einsatz von Beobachtern für Konsens mit relativen Ausgangsmessungen (Wieland and Allgöwer, 2010). Es werden die grundsätzlichen Herausforderungen diskutiert, die sich beim Beobachterentwurf in diesem Zusammenhang ergeben, und es wird ein Lösungsansatz vorgeschlagen, der diesen Herausforderungen adequat begegnet, indem das Problem auf ein Zustandsschätzproblem mit unbekannten Eingängen zurückgeführt wird. Es werden Lösungen vorgeschlagen, die auf Zustandsschätzern mit voller und reduzierter Ordnung und mit unbekannten Eingängen basieren. Der Schätzer mit reduzierter Ordnung stellt dabei geringere Anforderungen an die einzelnen Systeme, macht jedoch Anpassungen der Methoden aus dem vorhergehenden Kapitel nötig.

Kapitel 5 – The Internal Model Principle for Consensus and Synchronization (Das Innere-Modell-Prinzip für Konsens und Synchronisation)

Während sich die vorhergehenden Kapitel mit Konsensproblemen in homogenen Netzwerken beschäftigt haben, d.h. in Netzwerken, in denen die dynamischen Modelle aller einzelner Systeme identisch sind, wird in diesem Kapitel der Frage nachgegangen, unter welchen Voraussetzungen Konsens und Synchronisation in heterogenen Netzwerken möglich ist (Wieland and Allgöwer, 2009a,b, Wieland et al., 2010c). Die Beantwortung dieser Frage führt sowohl in linearen als auch in nichtlinearen Netzwerken auf das Innere-Modell-Prinzip für Konsens und Synchronisation. Dieses Prinzip besagt, dass alle einzelnen Systeme, ggf. gemeinsam mit einem lokalen dynamischen Regler, ein internes Modell eines gemeinsamen virtuellen Exosystems enthalten müssen. Die Bedingungen an die einzelnen Systeme entsprechen dabei denjenigen, die von Problemen der linearen und nichtlinearen Ausgangsfolgeregelung als Francis-Gleichungen bzw. Francis-Byrnes-Isidori-Gleichungen bekannt sind. Dadurch werden zwei wichtige Gebiete der System- und Regelungstheorie, nämlich Konsens und Synchronisation einerseits und Folgeregelung andererseits, in einen Zusammenhang gesetzt.

Kapitel 6 – Synchronization in Heterogeneous Networks with Dynamic Couplings (Synchronisation in heterogenen Netzwerken mit dynamischen Kopplungen)
In diesem Kapitel werden schließlich zwei neue Kopplungsmethoden vorgestellt. Die erste Methode führt zu Synchronisation in beliebigen linearen heterogenen Netzwerken sofern die einzelnen Systeme entdeckbar und stabilisierbar sind (Wieland et al., 2010c). Die zweite Methode führt zu Synchronisation in Netzwerken von unterschiedlichen nichtlinearen Oszillatoren (Wieland et al., 2010b). Beide Methoden nutzen die Erkenntnisse aus dem vorhergehenden Kapitel. Die dynamischen Kopplungen werden so gewählt, dass durch einen Vorsteuerungsanteil das Innere-Modell-Prinzip erfüllt wird und ggf. gemeinsam mit einem stabilisierenden Rückführungsanteil Synchronisation garantiert wird. Für das lineare Netzwerk wird der stabilisierende Rückführungsanteil analog zu einer Ausgangsfolgeregelung entworfen. Für das Netzwerk aus Oszillatoren stellt sich heraus dass keine stabilisierende Rückführung notwendig ist, da die strukturelle Robustheit exponentiell stabiler Grenzzyklen gemeinsam mit dem Vorsteuerungsanteil ausreicht um Synchronisation zu erreichen. Es wird schließlich gezeigt, dass die vorgeschlagenen dynamischen Kopplungen auch in homogenen Netzwerken höhere Systemkomplexität und höhere topologische Komplexität erlauben.

Chapter 1

Introduction

Consensus and synchronization are everywhere!

Admittedly, a very strong statement to start with. Most probably one of the strongest to be made in this thesis. As a matter of fact, there will be no stringent justification of this claim. The objective of the present thesis is not to provide a proof of the above statement. The reason to start with this striking claim is the fact that its apparent truth served as inspiration and motivation for the present work at large. *The contribution of this thesis is an improved understanding – in a systems theoretic framework – of some of the diverse mechanisms leading to phenomena summarized under the terms consensus and synchronization.* In particular, we will be interested in the question how the use of *static* consensus and synchronization mechanisms is a limiting factor in such phenomena and how mechanisms that employ *dynamic* couplings may remove some of those limitations.

At first, we will however substantiate the introductory statement. To this end, we first need to understand the meaning of the terms *consensus* and *synchronization* and how the two relate. Therefore, we will have a look on the two terms from an etymological perspective.

The word *consensus* derives from Latin *com* meaning 'with' or 'together with' and *sentire* meaning 'to feel'. Hence, consensus means to feel together with other individuals or to *agree* on a common feeling about something.

The term *synchronization* originates from Greek *syn* having the same meaning as *com* in consensus, namely 'with' or 'together with', and *chronos* meaning 'time'. Therefore, synchronization refers to the fact that individuals share a common notion of time or achieve *temporal coincidence* of some events.

From the above explanations, it is apparent, that the two terms are closely related. Both describe the effect of reaching agreement in a group of individuals in some sense. In the case of consensus, the relevant quantity is frequently the concrete value agreed upon. In contrast, in synchronization the focus is on a common timing. Therefore, consensus deals with the problem of agreement about the value of some variable while synchronization deals with the problem of agreement on the time instants, when a variable takes specific values. That said, one can of course argue that consensus about time instants yields synchronization, and synchronization of trajectories yields consensus. As a result, whether some phenomenon is termed consensus or synchronization often depends on the point of view one decides to take.

With this first systems theoretic picture of consensus and synchronization in mind, it is easy to find various examples of phenomena fundamentally relying on these principles in all kind of scientific disciplines. It is impossible to give an exhaustive list of such phenomena here. We only mention some examples from a range of scientific and engineering disciplines to illustrate how widespread examples of consensus and synchronization are. Many more examples are found in Pikovsky et al. (2001), Strogatz (2003).

Probably closest to the habitual language use of the word 'consensus' is the example of collaborative decision making (Baron et al., 1992). But consensus is also essential in other social phenomena like for example crowd dynamics (Helbing et al., 2000), where consensus mechanisms appear to play a crucial role in escape scenarios. Of course, examples in the living world are not restricted to humans. Flocking or swarming (see Reynolds, 1987, Tanner et al., 2007) are extremely fascinating phenomena to be observed, e.g., in groups of birds or fish. There seems to be a spontaneous agreement among all individuals in the group about the new direction. Such phenomena keep exciting researchers and eliciting the quest for scientific explanations. Another example from biology, which keeps astonishing researchers since more than two centuries, is the phenomenon of fireflies flashing in synchrony, to be observed, e.g., in Southeast Asia (Buck, 1938, 1988). Besides those fascinating inter-organism examples, consensus and synchronization play a vital role within single organisms. Neuroscientists believe that synchronization is responsible for phenomena like, e.g., epilepsy, locomotion, or particular rhythms observed in electroencephalography (see Izhikevich, 2007).

In the engineering world, one of the most commonly cited examples for consensus and synchronization is the problem of coordinated motion of individual mobile agents in its various occurrences. Examples include Fax and Murray (2002, 2004), Jadbabaie et al. (2003), Nair and Leonard (2008), Olfati-Saber (2006), Qu et al. (2008), Ren et al. (2007), Sepulchre et al. (2008), Tanner et al. (2003). But the range of examples of consensus and synchronization reaches way beyond coordinated motion. Recently, consensus and synchronization have been used, e.g., to control a very large telescope (Sarlette et al., 2010), for force allocation in a paper moving device (Fromherz and Jackson, 2003), to control the directional sensitivity of smart antennas (Hutu et al., 2009), or for speed synchronization in motors (Zhao et al., 2010), to name only a few.

The above list of examples illustrates, that consensus and synchronization is relevant in an extremely wide range of applications from various disciplines including, but not limited to, social sciences, biology, physics, and engineering. This observation is supported by Pikovsky et al. (2001) and justifies the statement at the outset of this chapter: indeed *consensus and synchronization are everywhere*!

1.1 Dimensions of Complexity in Consensus and Synchronization

In this thesis, we are taking a systems theoretic point of view on consensus and synchronization in groups of individuals. In this context, individuals are commonly modeled as dynamical systems usually described by ordinary differential equations (ODEs). Those individuals are said to reach consensus or to synchronize if they get close to each other in the sense that some specific quantities, like outputs or states, of all individual systems are asymptotically identical, or at least close to each other. To achieve this goal, it is necessary that all individuals in the group share some amount of common information,

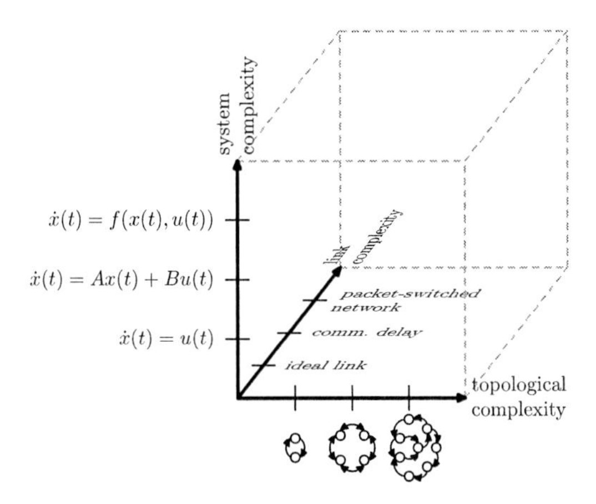

Figure 1.1: The complexity cube for coupled dynamical systems.

i.e., the individuals in the group are coupled in some way.

A naive approach to tackle consensus and synchronization problems would be to consider the whole group together with the couplings as a single, very large dynamical system. This corresponds to some extent to the classical point of view in systems and control theory, where the objects to analyze and to design controllers for are *single*, potentially large dynamical systems. This point of view however neglects that, in many instances, large dynamical systems exhibit highly structured behavior (see Lin, 2006) and admit a natural decomposition in smaller subsystems with a specific interconnections (see Vidyasagar, 1981). To cope with large, interconnected systems in a scalable[1] fashion and in response to continuously increasing system complexity, it is indispensable to take the structure of those systems into account for both analysis and design.

Consensus and synchronization of coupled individual systems are often considered to represent canonical examples of such structured behavior. If we assume couplings between the individual systems to be described by some specific interconnection topology, three independent dimensions of complexity can be identified: complexity of individual system dynamics, complexity of the interconnection topology, and complexity of the individual links that actually couple the systems. These three dimensions of complexity are illustrated in the *complexity cube* in Figure 1.1. System complexity may range, e.g., from simple linear integrator models to complex nonlinear systems or hybrid systems. Topological complexity deals with the question 'who talks to whom' or 'who listens to whom' and typically ranges from single feedback loops to complex, time-varying interconnection topologies, where individuals may join or leave the group at any time. The couplings between the individual systems regularly stem from some level of communication, usu-

[1]In the context of interconnected dynamical systems, *scalability* is the property that analysis and design complexity grows slowly as compared to system size, i.e., number of subsystems. Usually, the term scalability is used in software engineering and describes the property of software systems to handle growing amounts of work or system size in a graceful manner (Hill, 1990).

ally subject to communication constraints like, e.g., limited number of links or limited sensing ranges. Severe communication constraints yield high topological complexity and weak communication constraints yield low topological complexity. We will use the terms interconnection topology and communication topology interchangeably in what follows. Link complexity deals with the question how sending information across the network influences the information. It takes into account that the coupling links between individual systems are imperfect, e.g., because they are subject to time delays (see Olfati-Saber and Murray, 2004, Papachristodoulou et al., 2010, Sun et al., 2008) or suffering from effects of packet-switched networks (see Hespanha et al., 2007, Olfati-Saber et al., 2007, Zampieri, 2008). Problems arising along the latter dimension of complexity are often summarized under the term *networked control systems.* These problems are not subject of the present work, i.e., we assume all links to be perfect and focus our attention to system complexity and topological complexity.

These two remaining dimensions of complexity help to make a distinction between consensus problems and synchronization problems, as they are commonly considered in the context of dynamical systems. Even though, the definitions of the two terms lack clear distinctive features, the common usage of the terms 'consensus' and 'synchronization' applies to rather different problems. In what follows, we will discuss the difference between those problems in terms of the dimensions of complexity introduced above.

Consensus research is mostly focused on topological complexity, in particular on communication constraints, while individual system dynamics are usually fairly simple. Commonly, individual systems are modeled as simple integrators (Jadbabaie et al., 2003, Lee and Spong, 2007) and the dynamic evolution of the group is entirely determined by the couplings, i.e., the exchange of information between the individuals (see Carli et al., 2008, Cortés, 2008, Fax and Murray, 2004, Goldman and Zilberstein, 2003, Jadbabaie et al., 2002, 2003, Moreau, 2004b, 2005, Ren, 2007, Ren and Beard, 2005, Tahbaz-Salehi and Jadbabaie, 2008, Wang et al., 2008, and the references therein). A common extension, which is still an active field of research, is to consider individuals modeled as chains of integrators (Hu and Lin, 2010, Ren and Atkins, 2005, Ren et al., 2006).

In contrast to consensus, the topic of *synchronization* usually deals with more complex system dynamics – for instance nonlinear oscillators – while putting less emphasis on the topological complexity. Often, the extreme case of an all-to-all coupling is assumed, which can be considered as unconstrained communication. The problem of synchronization was first addressed by Huygens in 1665 (see Bennett et al., 2002). More recent examples can be found in Canavier and Achuthan (2007), Hoppensteadt and Izhikevich (1997), Kopell and Ermentrout (1986), Slotine and Wang (2004), Stan and Sepulchre (2007), Wang and Slotine (2005), Wu (2002).

The observation that consensus and synchronization research often focuses on either topological or system complexity, but rarely both at the same time, raises two important questions: What are the reasons for this tradeoff between system complexity and topological complexity in existing results and how can we get over this limitation? These questions will be partly answered in the present thesis. In particular, we will motivate and illustrate the use of dynamic couplings, i.e., couplings that are extended by dynamic compensators, to achieve consensus and synchronization in networks of systems with elevated complexity, including non-identical system dynamics, and couplings that are subject to strong communication constraints.

Concluding this section, a remark on other relevant types of structured behavior is indicated. For instance, Reynolds (1987) identifies three rules that control flocking behavior, namely *alignment*, *cohesion*, and *separation*. While the first two are closely related to consensus and synchronization, the opposite seems to be true for the latter one. In this case, the structured behavior can be interpreted as cooperative constraint satisfaction (see Bürger and Guay, 2008, Chang et al., 2003, Wieland and Allgöwer, 2007, Wieland et al., 2007, and the references therein). The scope of the present thesis is however focused on the behavior described above, i.e., consensus and synchronization.

1.2 Coupling Mechanisms

In the discussion above, we have not been specific about the mechanisms used to actually realize the couplings between the individual systems in the network. Among the various mechanisms potentially leading to consensus and synchronization, the one which is probably most commonly used is *diffusive couplings* (Hale, 1997).

We will give two examples to illustrate the idea. A publication that served as inspiration for many and triggered the interest in consensus of researchers in systems and control is due to Tamás Vicsek (Vicsek et al., 1995). He considered the problem of N particles moving in the plane with a common, constant speed and was interested in update schemes for the headings of the individual particles leading to alignment. The update scheme he proposed, which is commonly referred to as the *Vicsek model*, has been interpreted as a feedback mechanism in Jadbabaie et al. (2002, 2003). It can be defined as

$$\theta_k(t+1) - \theta_k(t) = \Delta_k(t) + u_k(t) \tag{1.1a}$$

$$u_k(t) = -\frac{1}{1+|\mathcal{N}_k(t)|}\sum_{j\in\mathcal{N}_k(t)} \left(\theta_k(t) - \theta_j(t)\right), \tag{1.1b}$$

where $\theta_k(t) \in \mathbb{S}^1$ is the heading of the kth particle at time t, $\mathcal{N}_k(t) \subseteq \{1,\dots,N\}$ is the set of indices of the neighbors of the kth particle at time t, i.e., describes the interconnection topology, and $\Delta_k(t) \in \mathbb{S}^1$ represents noise for all $k \in \mathbb{N}_N$.

The Vicsek model has a striking similarity (Sepulchre et al., 2004) to a famous model for synchronization, namely the *Kuramoto model* (Acebrón et al., 2005, Kuramoto, 1975)

$$\dot{\phi}_k(t) = \omega_k + u_k(t) \tag{1.2a}$$

$$u_k(t) = -\sum_{j=1}^{N} K_{k,j}\sin\left(\phi_k(t) - \phi_j(t)\right), \tag{1.2b}$$

where $\phi_k(t) \in \mathbb{S}^1$ is the kth oscillator's phase, $\omega_k \in \mathbb{R}$ is the kth oscillator's natural frequency, and $K_{k,j} \in \mathbb{R}$ are coupling gains for $k \in \mathbb{N}_N$.

The couplings (1.2b) in the Kuramoto model can be viewed as a nonlinear version of the linear couplings (1.1b) of the Vicsek model. In fact, the nonlinearity in (1.2b) takes into account the symmetry of the configuration space $\mathbb{S}^1$ of the Kuramoto model (see Sarlette, 2009, Sarlette and Sepulchre, 2009a,b). In both cases, the couplings are expressed in terms of *weighted sums of relative information*, which is the defining property of diffusive couplings. The diffusive couplings in the Vicsek and the Kuramoto model have the effect that the headings $\theta_k(t)$, $k \in \mathbb{N}_N$ and the phases $\phi_k(t)$, $k \in \mathbb{N}_N$ converge to an average of the headings respectively phases of the neighboring systems.

To make the term *diffusive couplings* plausible, consider a one-dimensional diffusion model $\frac{\partial \theta(t,z)}{\partial t} = \frac{\partial^2 \theta(t,z)}{\partial z^2}$ with time $t \in \mathbb{R}$ and space coordinate $z \in \mathbb{R}$. Define $\theta_k(t) \triangleq \theta(t, kh)$ and $\mathcal{N}_k \triangleq \{k-1, k+1\}$ for some $h \in \mathbb{R}_+$ and all $k \in \mathbb{N}$. Then we have

$$\dot{\theta}_k(t) = \left. \frac{\partial^2 \theta(t,z)}{\partial z^2} \right|_{z=kh} = -\frac{1}{h^2} \sum_{j \in \mathcal{N}_k} (\theta_k(t) - \theta_j(t)) + \mathcal{O}\left(h^2\right), \quad k \in \mathbb{N}.$$

Thus, the couplings (1.1b) in the Vicsek model can be viewed as a discretized diffusion model. Similar arguments hold for the nonlinear couplings (1.2b).

Diffusive couplings often seem a natural choice to achieve consensus or synchronization in groups of individual systems. They possess a number of beneficial properties. Namely, diffusive couplings can be realized using *relative sensors*, i.e., no absolute reference framework is required. In addition, diffusive couplings represent a completely *decentralized* mechanism while still allowing for certain systems in the group to act as leaders or reference by an appropriate choice of the communication topology. Furthermore, once synchronization is achieved, diffusive couplings vanish and the individuals evolve independently. This fact has important implications that will be discussed in detail in this thesis.

Note that the diffusive couplings used in the Vicsek model and the Kuramoto model are *static couplings* in the sense that they only depend on the instantaneous value of relative states and do not possess a coupling state. The system inputs are related to the relative states by an algebraic equation. As opposed to this, *dynamic couplings* are characterized by the property that they depend on the instantaneous value of the relative states and on a coupling state, i.e., they are themselves described by a dynamical system. In dynamic couplings, the system inputs are related to the relative states by a differential equation. This corresponds to static as opposed to dynamic feedback in classical feedback control.

In view of the above discussion, we will analyze static diffusive couplings among linear systems in the first part of this thesis. We will discuss some existing and provide several new extensions to the Vicsek model. Graph theory, in particular *algebraic graph theory*, will be used to model the communication topology in the network, and graph properties will be related to properties of the interconnected network of dynamical systems. It will turn out that the use of static diffusive couplings requires strong assumptions on the individual systems and the communication topology. Therefore, we show in the second part of this thesis, how to extend static diffusive couplings by adding coupling *dynamics* in order to be able to relax parts of those assumptions. First, we discuss output consensus with static diffusive couplings extended by dynamic observers under the constraint of relative output sensing. Afterwards, we consider consensus and synchronization in heterogeneous networks, i.e., networks of individual systems that possess non-identical dynamical models. We will give necessary conditions for synchronization in terms of an internal model requirement for linear and nonlinear networks and provide constructive methods to synchronize networks of general non-identical linear systems and networks of non-identical nonlinear oscillators.

1.3 Outline and Contributions

The following overview reveals the outline of this thesis and briefly summarizes its contributions.

Chapter 2 – Graph Theory for Consensus Problems
In this chapter we review basic definitions from graph theory and provide new interpretations linking graph theory to consensus problems. Parts of this chapter are based on Wieland et al. (2008, 2010a).

- We review basic definitions of graph theory, as far as they are relevant for consensus and synchronization problems. In particular, we introduce relevant definitions of graph connectedness.
- We establish the link between algebraic graph theory, namely properties of the Laplacian matrix and the adjacency matrix, and diffusive couplings.
- We give new interpretations of the eigenstructure of the graph Laplacian, introducing the concept of isolated strongly connected components as an extension of connected components of undirected graphs to directed graphs.

Chapter 3 – Consensus with Static Diffusive Couplings
This chapter addresses different algorithms that employ static diffusive couplings to achieve consensus among identical linear systems over fixed and time-varying communication graphs. In particular we discuss interdependencies between system properties and topological properties of the network. Parts of this chapter are based on Wieland et al. (2008, 2010a).

- We discuss and partly improve existing results for consensus among identical linear systems with static diffusive couplings.
- A new LMI-based design method is proposed to determine the gains to be used in static diffusive couplings.
- We propose methods to achieve consensus in a network with a leader, potentially with different dynamics than the remaining group. The results give conditions on the leader dynamics and explain how to connect the leader to the existing group.

Chapter 4 – Observer-Based Output Consensus with Relative Output Sensing
This chapter identifies the specific difficulties induced by relative sensing in the design of observers to be used in conjunction with static diffusive couplings and proposes a method to circumvent these difficulties. The results in this chapter are available in Wieland and Allgöwer (2010).

- We explain why unknown-input observers need to be used in conjunction with static diffusive couplings and relative sensing and why time-varying communication topologies induce additional problems.

- We propose a design method to implement static diffusive couplings with full order unknown-input observers.
- We relax the constraints imposed by a full order unknown input observer by proposing a design based on reduced order unknown input observers.

Chapter 5 – The Internal Model Principle for Consensus and Synchronization
In this chapter, we deal with the question of consensus and synchronization in heterogeneous networks and provide new necessary conditions for consentability and synchronizability. This chapter is mainly based on Wieland and Allgöwer (2009a,b), Wieland et al. (2010c).

- We derive necessary conditions for consentability and synchronizability of heterogeneous linear networks in terms of linear matrix equations.
- We derive similar necessary conditions for consentability and synchronizability of heterogeneous nonlinear networks in terms of nonlinear partial differential equations.
- The linear and nonlinear conditions are interpreted in terms of an internal model requirement.
- We establish a close link between the necessary conditions for consentability and synchronizability and conditions known from linear and nonlinear output regulation.

Chapter 6 – Synchronization in Heterogeneous Networks with Dynamic Couplings
This chapter proposes new mechanisms for synchronization in heterogeneous networks based on the internal model requirement from the previous chapter. It is shown that coupling dynamics are necessary to solve the problem. Most of this chapter is based on Wieland et al. (2010b,c).

- We show, by constructing appropriate dynamic coupling controllers, that mild technical assumptions are sufficient to make the necessary conditions from the previous chapter also sufficient for synchronizability of heterogeneous linear networks.
- We derive a method, based on entrainment of nonlinear oscillators by weak forcing, that allows to satisfy the internal model requirement from the previous chapter in groups of non-identical nonlinear oscillators and show that this is sufficient to achieve synchronization.

Chapter 7 – Conclusions
This final chapter provides some conclusive remarks summarizing the thesis and hints to possible future directions of research.

Supplementary material is provided in several appendices, referenced at appropriate places, with the aim to make this thesis self-contained.

Chapter 2

Graph Theory for Consensus Problems

In the study of consensus and synchronization problems, we deal with networks of individual dynamical systems that are interconnected according to a specific *communication topology*. The communication topology characterizes the links between the individual systems, i.e., it describes which individuals are close to each other within the network. More precisely, the communication topology determines the set of neighbors – or predecessors and successors if links are directional – for every member of the network, which is certainly a topological property.

In this context, the term *communication* is to be understood in a very broad and general sense. Specifically, the meaning of communication is not restricted to exchange of information via some high-level network that possesses the ability to transmit arbitrary information like, e.g., a digital network, but communication may take place for instance by means of physical links or some level of relative sensing, to name only a few among a multitude of possible types of interactions among individual systems. Therefore, whenever we talk about communication topologies in this thesis, we do that without being specific about the method used to interconnect individuals in the network.

To suitably model the communication topology in a network of individual systems, graph theory proved to be a useful tool (see Fax and Murray, 2002, de Gennaro and Jadbabaie, 2006, Jadbabaie et al., 2003, Kim and Mesbahi, 2006, Wieland et al., 2008, 2010a). Graph theory, and more specifically *algebraic* graph theory, is particularly useful in the case of diffusively coupled systems (as introduced in Section 1.2) as will be demonstrated in this chapter. Early results on algebraic graph theory, which related spectral properties of specific matrices to graph properties, can be found in von Collatz and Sinogowitz (1957), Fiedler (1973, 1975). In particular, the seminal papers by Miroslav Fiedler introduce among other things the *algebraic connectivity* of graphs which proved to be a very powerful tool since the time it was first published. Since then, graph theory has been and remains to be a very active field of research. See Bollobás (1998), Diestel (2005), Godsil and Royle (2004), Merris (1995), Mohar (1992), Newman (2000), Wu (2005), and the references therein for an overview over the topic of graph theory.

In this chapter, we summarize the most important definitions and results from graph theory and algebraic graph theory as far as they are relevant for the study of consensus problems. Furthermore, we generalize the concept of *connected components* of undirected graphs to directed graphs.

2.1 Basic Definitions and Link to Diffusive Couplings

Graphs consist of *vertices* (or nodes) and *edges* (or arcs) connecting the vertices. Edges can be undirected or directed. The latter case results in a directed graph, abbreviated *digraph*. In addition, weights can be assigned to the (directed) edges to obtain a *weighted (di)graph*.

In this thesis, we consider the most general case of weighted digraphs – containing undirected and unweighted graphs as special cases – to model the communication topology of networks of interconnected systems. In this context, it is often required to consider the communication topology as a network characteristic which may change over time. For that reason, it is appropriate to allow the graph to depend on time $t \in \mathbb{R}$.

Definition 2.1 (Communication Graph). *A time-varying communication graph $\mathcal{G}(t)$ is a tuple $\mathcal{G}(t) = \{\mathcal{V}, \mathcal{E}(t), W(t)\}$ of vertices $\mathcal{V} = \{v_1, \ldots, v_N\}$ with $|\mathcal{V}| = N > 0$, edges $\mathcal{E}(t) \subset \mathcal{V} \times \mathcal{V}$, and adjacency matrix $W(t) = [w_{k,j}(t)] \in \overline{\mathbb{R}}_+^{N\times N}$, where $t \in \mathbb{R}$ represents time, satisfying the following properties:*

(P1) The graph contains no self-loops, i.e., $(v_k, v_k) \notin \mathcal{E}(t)$ and $w_{k,k}(t) = 0$ for all $t \in \mathbb{R}$ and all $k \in \mathbb{N}_N$.

(P2) The elements $w_{k,j}(\cdot) : \mathbb{R} \to \overline{\mathbb{R}}_+$, $k, j \in \mathbb{N}_N$ of the adjacency matrix W are non-negative, piecewise continuous[1], and bounded functions of time.

(P3) Given $k, j \in \mathbb{N}_N$, the edge (v_j, v_k) is contained in $\mathcal{E}(t)$ at time t if and only if $w_{k,j}(t) > \alpha$ for some fixed threshold $\alpha \in \mathbb{R}_+$.

In the above definition, the vertices v_k, $k \in \mathbb{N}_N$ represent the individual systems and the edges $(v_j, v_k) \in \mathcal{E}(t)$ model interconnections between the individual systems. For a given edge $(v_j, v_k) \in \mathcal{E}(t)$, we denote the first vertex v_j as the *tail* and the second vertex v_k as the *head* of the edge e and adopt the convention that *information* flows from the tail to the head of an edge, i.e., the system represented by the tail of an edge influences the system represented by the head of that edge. Usually, edges are represented by arrows pointing from the tail to the head. An example of a weighted digraph with five vertices, with edges represented by arrows, is depicted in Figure 2.1.

The elements $w_{k,j}(t)$, $k, j \in \mathbb{N}_N$ of the adjacency matrix $W(t)$ quantify the weights for the corresponding edges $(v_j, v_k) \in \mathcal{E}(t)$ at time t. It should be remarked that representations of the adjacency matrix $W(t)$ depend on a specific ordering of the vertices, i.e., the adjacency matrix for a given graph $\mathcal{G}(t)$ is uniquely defined modulo vertex permutations. Of course, spectral properties of $W(t)$ are invariant with respect to these permutations and the results in this chapter do not depend on a specific vertex ordering. Properties (P2) and (P3) from Definition 2.1 restrict the evolution of the edge weights $w_{k,j}(\cdot)$, $k, j \in \mathbb{N}_N$ over time. Property (P2), besides imposing piecewise continuity, excludes the case that edge weights grow unbounded over time, while Property (P3) excludes the case that edge weights become arbitrarily small by imposing α as a uniform lower bound. More precisely,

[1] A function $w : \mathbb{R} \to \mathbb{R}$ is said to be piecewise continuous if it has at most finitely many points of discontinuity in any closed interval of its domain and at the points of discontinuity the left and right limits exist (see Lang, 1987).

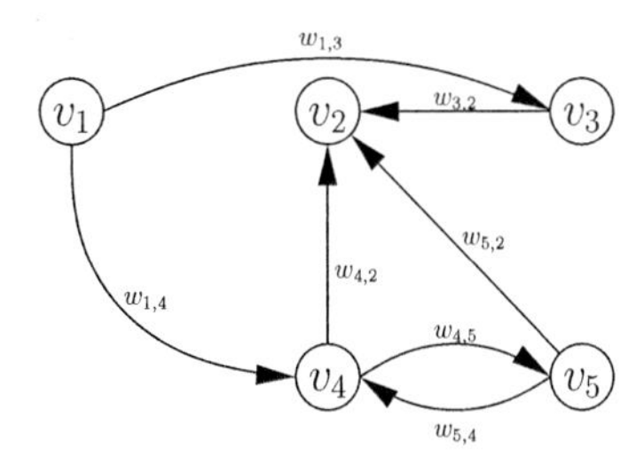

Figure 2.1: Weighted digraph with five vertices.

the values $w_{k,j}(t)$ may get arbitrarily close to zero at specific times t or for $t \to \infty$, but the corresponding edge will not be present in the graph if $w_{k,j}(t) \leq \alpha$. The requirements imposed by Properties (P2) and (P3) are actually only relevant for time-varying graphs, in which case fixed upper and lower bounds must exist uniformly in t. Thereby, the precise values of these bounds, in particular the threshold α, is not important and will never be used. For fixed times t, i.e., for fixed graphs, upper and lower bounds for the edge weights always exist. Therefore, when dealing with fixed graphs, we will frequently and without loss of generality impose the following assumption:

Assumption 2.2. *The graph $\mathcal{G}(t) = \{\mathcal{V}, \mathcal{E}(t), W(t)\}$ is such that the elements of the adjacency matrix $W(t)$ satisfy $w_{k,j}(t) \in \{0\} \bigcup (\alpha, \infty) \subset \overline{\mathbb{R}}_+$ for some fixed threshold $\alpha \in \mathbb{R}_+$, i.e., $(v_j, v_k) \in \mathcal{E}(t)$ if and only if $w_{k,j}(t) > 0$ for all $k, j \in \mathbb{N}_{|\mathcal{V}|}$.*

It should be noted that, for a fixed set of vertices $\mathcal{V}$ and a fixed threshold α, the graph $\mathcal{G}(t)$ is entirely characterized by the adjacency matrix $W(t)$ by Property (P3). Therefore, we shall write $\mathcal{E}(W(t)) \subset \mathcal{V} \times \mathcal{V}$ to denote the edge set induced by some adjacency matrix $W(t)$ and $\mathcal{G}(W(t)) = \{\mathcal{V}, \mathcal{E}(W(t)), W(t)\}$ to denote the graph induced by that matrix.

Some special cases of graphs are included in Definition 2.1 as follows: A graph $\mathcal{G}(t)$ is *unweighted* if the elements of the adjacency matrix $W(t)$ are constraint to take values in $\{0, 1\}$. In that case, we may drop the adjacency matrix $W(t)$ from the graph description and simply write $\mathcal{G}(t) = \{\mathcal{V}, \mathcal{E}(t)\}$. The special case of an *undirected* graph $\mathcal{G}(t)$ is obtained by imposing $w_{k,j}(t) = w_{j,k}(t)$ for all $k, j \in \mathbb{N}_N$, i.e., $W(t) = W^T(t)$. This implies that $(v_j, v_k) \in \mathcal{E}(t)$ if and only if $(v_k, v_j) \in \mathcal{E}(t)$ for all $k, j \in \mathbb{N}_N$. Another important class of weighted digraphs worth mentioning is the class of balanced graphs. A graph $\mathcal{G}(t)$ is *balanced* if $\sum_{j=1}^{N} w_{k,j} = \sum_{j=1}^{N} w_{j,k}$ for all $k \in \mathbb{N}_N$, i.e., the sum of the edge weights of the incoming edges equals the sum of the edge weights of the outgoing edges. Balanced graphs enjoy many of the properties of undirected graphs while the restriction to balanced graphs is substantially weaker than requiring the graph to be undirected. Since in undirected graphs the adjacency matrix is symmetric, any undirected graph is also a balanced graph.

Later on, we will need the concept of (unweighted) *induced subgraphs*. Given a graph $\mathcal{G}(t) = \{\mathcal{V}, \mathcal{E}(t), W(t)\}$. Consider a subset $\tilde{\mathcal{V}} \subseteq \mathcal{V}$ of the vertices of $\mathcal{G}(t)$ with $|\tilde{\mathcal{V}}| = \tilde{N}$. The subgraph $\tilde{\mathcal{G}}(t)$ induced by $\tilde{\mathcal{V}}$ is the graph $\tilde{\mathcal{G}}(t) = \{\tilde{\mathcal{V}}, \tilde{\mathcal{E}}(t)\}$, where $\tilde{\mathcal{E}}(t) \triangleq \{(v, w) \in \mathcal{E}(t) : v, w \in \tilde{\mathcal{V}}\}$. That is, the induced subgraph $\tilde{\mathcal{G}}(t)$ is obtained from $\mathcal{G}(t)$ by removing some of the vertices together with all edges starting or ending at these vertices.

2.1.1 Graph Connectedness

In order to achieve a consensus among individual systems, a basic requirement is that all individuals share a minimum amount of common information. As explained before, information propagates in the network along the (directed) edges of the graph modeling the communication topology. To ensure that common pieces of information may reach all individuals, the graph needs to be connected in some appropriate sense. To introduce different definitions of connectivity, we first need to define a path in a graph as follows:

Definition 2.3 (Path). *For fixed t, a* path *p from vertex v_{l_1} to vertex v_{l_r} in a graph $\mathcal{G}(t)$ is a sequence $\{v_{l_1}, \dots, v_{l_r}\}$ of $r > 1$ distinct vertices such that $(v_{l_i}, v_{l_{i+1}}) \in \mathcal{E}(t)$, $i = 1, \dots, r-1$. The length of the path is $r - 1$.*

Existence of a path p from vertex $v \in \mathcal{V}$ to vertex $w \in \mathcal{V}$ in some graph $\mathcal{G}(t)$ implies that information can propagate from the system represented by vertex v to the system represented by vertex w. With the help of paths, we can define the set of vertices which can receive information from a specific vertex $v \in \mathcal{V}$, called the set of *descendants* of v, as

$$\mathcal{D}(v, \mathcal{G}(t)) \triangleq \{w \in \mathcal{V}, \exists \text{ path from } v \text{ to } w \text{ in } \mathcal{G}(t) \text{ at time } t\}.$$

The property of being descendant is transitive, i.e., $v_1 \in \mathcal{D}(v_2, \mathcal{G}(t))$ and $v_2 \in \mathcal{D}(v_3, \mathcal{G}(t))$ implies $v_1 \in \mathcal{D}(v_3, \mathcal{G}(t))$. If there is no ambiguity of the graph $\mathcal{G}(t)$, we may drop the second argument and simply write $\mathcal{D}(v)$ instead of $\mathcal{D}(v, \mathcal{G}(t))$ to denote the descendants of v in $\mathcal{G}(t)$. With these preliminaries at hand, we give the following definition of graph connectedness:

Definition 2.4 (Connected at Time t, Centroid). *A graph $\mathcal{G}(t) = \{\mathcal{V}, \mathcal{E}(t), W(t)\}$ is* connected at time t *if there exists a vertex $v \in \mathcal{V}$ such that for all $w \in \mathcal{V} \setminus \{v\}$, $w \in \mathcal{D}(v, \mathcal{G}(t))$, i.e., $\{v\} \bigcup \mathcal{D}(v, \mathcal{G}(t)) = \mathcal{V}$. The vertex v is called a* centroid *of $\mathcal{G}(t)$.*

If a graph $\mathcal{G}(t)$ is connected at time t, then its vertex set $\mathcal{V}$ contains at least one specific vertex v, called a centroid of $\mathcal{G}(t)$, from which information can propagate to all other vertices along paths in $\mathcal{G}(t)$. A graph obtained from $\mathcal{G}(t)$ by removing all edges that do not belong to one of theses paths is called a spanning tree of $\mathcal{G}(t)$. The centroid vertex v is the root of this spanning tree. Hence, a graph $\mathcal{G}(t)$ is connected at time t if a spanning tree exists at time t. It should be noted that neither the centroid vertex v nor the spanning tree need to be unique. Some authors refer to connected graphs in the sense of Definition 2.4 as *quasi strongly connected graphs* (Ren et al., 2007, Wu, 2005). The terminology used here is in accordance with Moreau (2005), Scardovi and Sepulchre (2009).

If a graph is connected, the information that may be shared throughout the network is determined by the information initially available at the centroid vertices. Sometimes, a stronger definition of connectedness is needed, in which information can propagate between any two vertices.

Definition 2.5 (Strongly Connected at Time t). *A graph $\mathcal{G}(t) = \{\mathcal{V}, \mathcal{E}(t), W(t)\}$ is* strongly connected at time t, *if any vertex $v \in \mathcal{V}$ is a centroid of $\mathcal{G}(t)$, i.e., for any pair of distinct vertices $v, w \in \mathcal{V}$, $w \in \mathcal{D}(v, \mathcal{G}(t))$ and $v \in \mathcal{D}(w, \mathcal{G}(t))$ or equivalently for any vertex $v \in \mathcal{V}$, $\{v\} \bigcup \mathcal{D}(v, \mathcal{G}(t)) = \mathcal{V}$.*

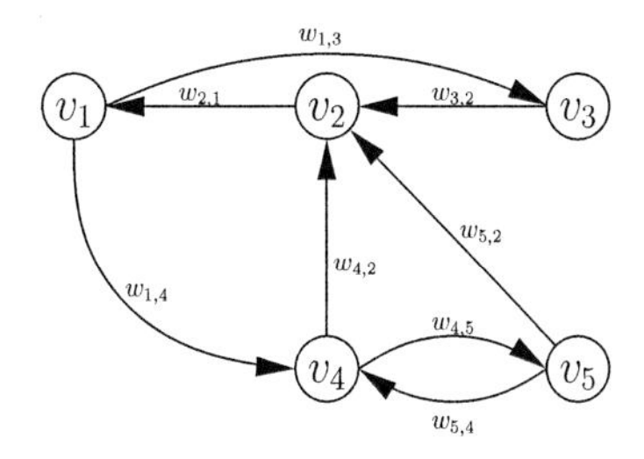

Figure 2.2: Strongly connected weighted digraph with five vertices.

To illustrate the above definitions, consider the graph depicted in Figure 2.1. This graph is connected with centroid v_1, but it is not strongly connected since v_1 is the only centroid. However, adding a single edge from vertex v_2 to vertex v_1, the resulting graph – depicted in Figure 2.2 – is strongly connected (and therefore also connected). To verify strong connectedness of the graph depicted in Figure 2.2, it suffices to find paths between all pairs of distinct vertices in the graph.

It is a remarkable fact that for balanced graphs the Definitions 2.4 of connectedness and 2.5 of strong connectedness are equivalent (Wu, 2005, Lemma 17). Since any undirected graph is balanced, the same holds of course true for undirected graphs.

Definitions 2.4 and 2.5 both deal with graph connectedness at a specific time t, i.e., for a fixed graph. Since the graph $\mathcal{G}(\cdot)$ may vary over time, definitions taking into account the evolution of the communication topology over time are of interest. To this end, we define the T-averaged adjacency matrix as

$$\overline{W}_T(t) = \frac{1}{T}\int_t^{t+T} W(\tau)\mathrm{d}\tau,$$

where $T \in \mathbb{R}_+$ is a time horizon. The graph induced by the averaged adjacency matrix, denoted $\overline{\mathcal{G}}_T(t) = \mathcal{G}(\overline{W}_T(t)) = \{\mathcal{V}, \overline{\mathcal{E}}_T(t), \overline{W}_T(t)\}$ with $\overline{\mathcal{E}}_T(t) = \mathcal{E}(\overline{W}_T(t))$, is called the union graph over intervals of length T. The definition below is taken from Moreau (2004b, 2005).

Definition 2.6 (Uniformly Connected). *A graph $\mathcal{G}(\cdot) = \{\mathcal{V}, \mathcal{E}(\cdot), W(\cdot)\}$ is called* uniformly connected, *if there exists a vertex $v \in \mathcal{V}$ and a time horizon $T \in \mathbb{R}_+$ such that v is a centroid of the union graph $\overline{\mathcal{G}}_T(t) = \mathcal{G}(\overline{W}_T(t))$ uniformly in t.*

It should be noted that uniformly connected is weaker than connected. In particular a graph $\mathcal{G}(\cdot)$ may be uniformly connected without ever being connected. Consider for example the two constant graphs $\mathcal{G}_1$ and $\mathcal{G}_2$ depicted in Figures 2.3(a) and 2.3(b). Both are not connected. Let $\mathcal{G}(t) = \mathcal{G}_1$ for $t \in [kT, (k+1/2)T)$ and $\mathcal{G}(t) = \mathcal{G}_2$ for $t \in [(k+1/2)T, (k+1)T)$ for all $k \in \mathbb{Z}$. Then the graph $\overline{\mathcal{G}}_T(t)$ is the constant, connected graph depicted in Figure 2.1, i.e., $\mathcal{G}(t)$ is uniformly connected without ever being connected.

Since any fixed, strongly connected graph is connected and any fixed, connected graph is uniformly connected, Definition 2.5 is the strongest of the three connectedness definitions

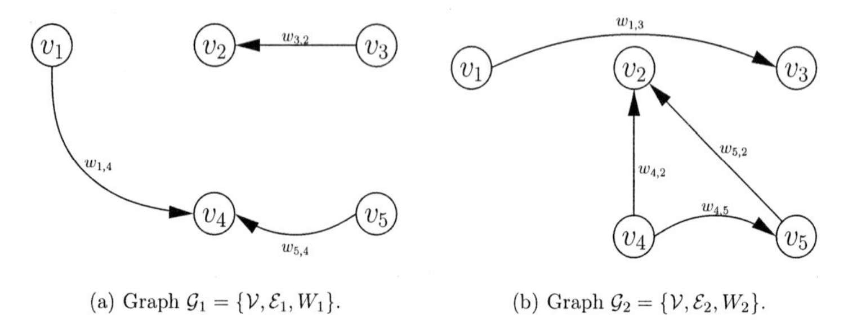

(a) Graph $\mathcal{G}_1 = \{\mathcal{V}, \mathcal{E}_1, W_1\}$.

(b) Graph $\mathcal{G}_2 = \{\mathcal{V}, \mathcal{E}_2, W_2\}$.

Figure 2.3: Two unconnected graphs $\mathcal{G}_1$ and $\mathcal{G}_2$ sharing a common vertex set $\mathcal{V}$ with connected union graph.

and Definition 2.6 is the weakest. In terms of complexity of the network (cf. Section 1.1), weaker connectedness assumptions imply less information propagated through the network or stronger communication constraints, i.e., higher topological complexity.

2.1.2 The Graph Laplacian

As a next step, we will establish a tight link between algebraic graph theory and diffusive couplings. To this end, consider again the Vicsek model (1.1) from Section 1.2 and assume the set of indices of the neighbors of the kth particle $\mathcal{N}_k(t)$ is defined by an unweighted digraph $\mathcal{G}(t) = \{\mathcal{V}, \mathcal{E}(t), W(t)\}$ with $|\mathcal{V}| = N$ vertices as $\mathcal{N}_k(t) = \{j \in \mathbb{N}_N : (v_j, v_k) \in \mathcal{E}(t)\}$. Then (1.1) can be rewritten as

$$\theta_k(t+1) - \theta_k(t) = \Delta_k(t) - \frac{1}{1 + \sum_{j=1}^{N} w_{k,j}(t)} \left(\sum_{j=1}^{N} w_{k,j}(t) \left(\theta_k(t) - \theta_j(t)\right) \right)$$

or, using the stacked vectors $\theta(t) = (\theta_1(t), \ldots, \theta_N(t))^T$ and $\Delta(t) = (\Delta_1(t), \ldots, \Delta_N(t))^T$,

$$\theta(t+1) - \theta(t) = \Delta - (I_N + D(t))^{-1} L(t) \theta(t). \tag{2.1}$$

In the latter expression, $D(t) \in \mathbb{R}^{N \times N}$ is the degree matrix of $\mathcal{G}(t)$, which is defined as $D(t) \triangleq \operatorname{diag}(W(t) 1_N)$, and $L(t) = [l_{k,j}(t)] \in \mathbb{R}^{N \times N}$ is the *graph Laplacian* matrix of $\mathcal{G}(t)$, defined as

$$L(t) \triangleq D(t) - W(t),$$

or element-wise

$$l_{k,j}(t) \triangleq \begin{cases} \sum_{i=1}^{N} w_{k,i}(t), & j = k, \\ -\, w_{k,j}(t), & j \neq k. \end{cases}$$

In (2.1), multiplication of the Laplacian matrix $L(t)$ by $(I_N + D(t))^{-1}$ from the left can be interpreted as normalization or additional weighting, which could also be included in the Laplacian matrix directly. The qualitative dynamic evolution of $\theta(t)$ is therefore determined by the Laplacian matrix $L(t)$. In particular, since $(I_N + D(t))^{-1}$ is non-singular,

$L(t)$ determines the equilibria of (2.1). Moreover, the graph Laplacian $L(t)$ uniquely determines the adjacency matrix $W(t)$ and therefore completely characterizes the graph $\mathcal{G}(t)$, just like the adjacency matrix. Hence, the graph Laplacian on the one hand characterizes the communication topology and on the other hand describes the interconnection dynamics for linearly diffusively coupled systems. This observation motivates the interest in relating algebraic properties of $L(t)$ to properties of the graph $\mathcal{G}(t)$, which is the main subject of the next section.

2.2 Algebraic Graph Properties in Consensus and Synchronization Problems

We start the discussion of algebraic graph properties by some general remarks on the spectrum of the Laplacian matrix $L(t)$. By definition, all rows of the Laplacian sum up to zero, diagonal elements are non-negative and off-diagonal elements are non-positive. These observations yield some important properties of the Laplacian (see Godsil and Royle, 2004):

The all ones vector 1_N is an eigenvector of $L(t)$ with corresponding eigenvalue 0 for all times $t \in \mathbb{R}$. In consensus problems this eigenvector spans the consensus subspace, i.e., the subspace where all individuals have reached a common value. The fact that the corresponding eigenvalue is 0 translates to the observation that diffusive couplings vanish once consensus is reached.

To determine the loci of the remaining eigenvalues of the Laplacian matrix, define $d_{\max}(t) \triangleq \max_{k \in \mathbb{N}_N} \sum_{j=1}^{N} w_{k,j}(t)$ be the maximum degree of a graph $\mathcal{G}(t)$. The Geršgorin Disk Theorem (Horn and Johnson, 1985, Theorem 6.1.1) implies that

$$\sigma(L(t)) \subset \{z \in \mathbb{C} : |z - d_{\max}(t)| \leq d_{\max}(t)\} \subset \overline{\mathbb{C}}_+,$$

i.e., the spectrum of the Laplacian is contained in a disk in $\mathbb{C}$ centered at $d_{\max}(t) + \mathrm{j}0$ with radius $d_{\max}(t)$ which is contained in the closed right-half complex plane. However, that disk depends on time t. To obtain a time-independent set containing the Laplacian eigenvalues, consider the supremum of the maximum edge weight $w_{\max} \triangleq \sup_{t \in \mathbb{R}} \max_{k,j \in \mathbb{N}_N} w_{k,j}(t)$. The supremum exists due to boundedness of the edge weights (cf. Property (P2) in Definition 2.1) and $d_{\max}(t) \leq (N-1)w_{\max}$ holds uniformly in t. Consequently, the Geršgorin Disk described above is uniformly contained in a disk in $\mathbb{C}$ centered at $(N-1)w_{\max} + \mathrm{j}0$ with radius $(N-1)w_{\max}$, i.e.,

$$\sigma(L(t)) \subset \{z \in \mathbb{C} : |z - (N-1)w_{\max}| \leq (N-1)w_{\max}\} \subset \overline{\mathbb{C}}_+$$

for all $t \in \mathbb{R}$. That is, all eigenvalues of the graph Laplacian are uniformly contained in a compact subset of the closed right-half complex plane $\overline{\mathbb{C}}_+$. The only intersection of this compact set with the imaginary axis $\mathrm{j}\mathbb{R}$ is the origin. As explained before, any Laplacian matrix has at least one eigenvalue at the origin with a particular interpretation in consensus problems. Below, we will have a closer look at this eigenvalue, first for undirected graphs and subsequently generalized to digraphs.

2.2.1 Connected Components and Algebraic Connectivity for Undirected Graphs

The graph property, which corresponds to the algebraic property represented by the kernel of the Laplacian matrix, is characterized by the connected components of an undirected graph. Connected components represent an important concept for undirected graphs, closely related to graph connectivity, as will be illustrated below.

Definition 2.7 (Connected Component). *At fixed time t, a* connected component *of an undirected graph $\mathcal{G}(t) = \{\mathcal{V}, \mathcal{E}(t), W(t)\}$ is an induced subgraph $\tilde{\mathcal{G}} = \{\tilde{\mathcal{V}}, \tilde{\mathcal{E}}\}$ which is maximal, subject to being connected. That is, $\tilde{\mathcal{G}}$ is connected and the unweighted graph induced by any set $\hat{\mathcal{V}}$ with $\tilde{\mathcal{V}} \subseteq \hat{\mathcal{V}} \subseteq \mathcal{V}$ is connected if and only if $\hat{\mathcal{V}} = \tilde{\mathcal{V}}$.*

As a consequence from the above definition, no path exists in $\mathcal{G}(t)$ between two vertices that belong to distinct connected components of $\mathcal{G}(t)$, i.e., each distinct connected component represents a subgraph which is independent of the remaining graph. Connected components are related to the graph Laplacian in the following lemma adopted from Fiedler (1973), Wu (2005):

Lemma 2.8. *Let t be fixed time and $\mathcal{G}(t) = \{\mathcal{V}, \mathcal{E}(t), W(t)\}$ an undirected graph with $|\mathcal{V}| = N$ vertices satisfying Assumption 2.2. Assume $\mathcal{G}(t)$ has $c \geq 1$ distinct connected components $\tilde{\mathcal{G}}_i = \{\tilde{\mathcal{V}}_i, \tilde{\mathcal{E}}_i\}$, $i \in \mathbb{N}_c$. Define vectors $p^i \in \mathbb{R}^N$, $i \in \mathbb{N}_c$ element-wise as*

$$p^i_j \triangleq \begin{cases} 1 & \text{if } v_j \in \tilde{\mathcal{V}}_i, \\ 0 & \text{if } v_j \notin \tilde{\mathcal{V}}_i, \end{cases} \qquad j \in \mathbb{N}_N.$$

Then, $\dim(\ker(L(t))) = c$ and $\ker(L(t)) = \ker(L^T(t)) = \operatorname{span}(p^1, \ldots, p^c)$, i.e., the c vectors p^i, $i \in \mathbb{N}_c$ span a non-negative, orthogonal basis of the kernel of the Laplacian matrix $L(t)$ (as well as the kernel of the transposed Laplacian matrix $L^T(t)$).

The proof of the above lemma is omitted, as it can be found in any standard book on algebraic graph theory (see Godsil and Royle, 2004). We only would like to mention that orthogonality of the vectors p^i, $i \in \mathbb{N}_c$ results from the fact that, at fixed time t, no vertex can belong to two distinct connected components. Since an undirected graph is connected if and only if it has exactly one connected component, namely the complete graph, and the Laplacian matrix $L(t)$ of an undirected graph is symmetric, connectivity implies that 0 is a simple eigenvalue of the Laplacian matrix $L(t)$. This yields the following corollary:

Corollary 2.9. *Let t be fixed time and $\mathcal{G}(t) = \{\mathcal{V}, \mathcal{E}(t), W(t)\}$ an undirected graph satisfying Assumption 2.2.*

Then $\mathcal{G}(t)$ is strongly connected (or equivalently connected) at time t if and only if the second smallest eigenvalue of the graph Laplacian $L(t)$ is positive, i.e. $\lambda_2(L(t)) > 0$.

The importance of the second smallest eigenvalue $\lambda_2(L(t))$ of the Laplacian matrix $L(t)$ was first recognized by Fiedler (1973, 1975), who introduced the term *algebraic connectivity* for that quantity. The remarkable properties of the algebraic connectivity of an undirected graph include the fact that its value cannot decrease when undirected edges are added to the graph. Therefore, the algebraic connectivity is a meaningful measure for the level of connectedness of an undirected graph.

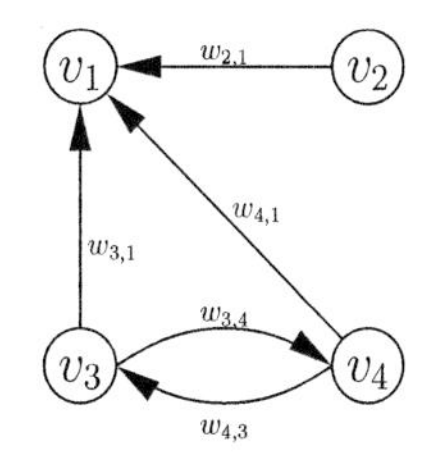

Figure 2.4: Directed graph with maximal subgraphs. The subgraphs induced by the vertex sets $\{v_1\}$, $\{v_2\}$, and $\{v_3, v_4\}$ are maximal subject to being strongly connected. The subgraphs induced by the vertex sets $\{v_1, v_2\}$ and $\{v_1, v_3, v_4\}$ are maximal subject to being connected.

2.2.2 Independent Strongly Connected Components for Directed Graphs

In undirected graphs, each vertex belongs to exactly one connected component, distinct connected components are mutually independent subgraphs, i.e., there exist no paths connecting them, and – due to equivalence of Definitions 2.4 and 2.5 – there is no ambiguity about the definition of a connected component. In directed graphs, the situation is more subtle. For a given graph $\mathcal{G}(t) = \{\mathcal{V}, \mathcal{E}(t), W(t)\}$, at fixed time t, there might be an induced subgraph $\tilde{\mathcal{G}} = \{\tilde{\mathcal{V}}, \tilde{\mathcal{E}}\}$ which is maximal subject to being strongly connected, but for which a path from some vertex $v \in \mathcal{V} \setminus \tilde{\mathcal{V}}$ to some vertex $\tilde{v} \in \tilde{\mathcal{V}}$, or the other way round, exists in $\mathcal{G}(t)$. In other words, induced subgraphs of directed graphs that are maximal, subject to being strongly connected, are not necessarily mutually independent. If one requires the induced subgraphs to be connected instead of strongly connected, distinct subgraphs, maximal subject to being connected, may even have vertex sets that are not disjoint. An example graph with subgraphs that are maximal subject to being connected and strongly connected respectively is depicted in Figure 2.4.

These properties, inherent to directed graphs, constitute a major difference to undirected graphs and explains why directed graphs pose additional challenges. The objective of this section is to introduce and explain a well-defined generalization of connected components of undirected graphs to digraphs. The particular choice taken in this thesis is motivated by the specific intended use for consensus problems.

Definition 2.10 (Independent Strongly Connected Component). *At fixed time t, an* independent strongly connected component *(iSCC) of a digraph $\mathcal{G}(t) = \{\mathcal{V}, \mathcal{E}(t), W(t)\}$ is an induced subgraph $\tilde{\mathcal{G}} = \{\tilde{\mathcal{V}}, \tilde{\mathcal{E}}\}$ which is maximal, subject to being strongly connected, and satisfies $(v, \tilde{v}) \notin \mathcal{E}(t)$ for any $v \in \mathcal{V} \setminus \tilde{\mathcal{V}}$ and $\tilde{v} \in \tilde{\mathcal{V}}$. That is $\tilde{G}$ is strongly connected and the unweighted digraph induced by any set $\hat{\mathcal{V}}$ with $\tilde{\mathcal{V}} \subseteq \hat{\mathcal{V}} \subseteq \mathcal{V}$ is strongly connected if and only if $\hat{\mathcal{V}} = \tilde{\mathcal{V}}$. Furthermore, there is no edge in $\mathcal{E}(t)$ with tail outside $\tilde{\mathcal{V}}$ and head in $\tilde{\mathcal{V}}$.*

The property, that no edge starting outside an iSCC and ending inside an iSCC may exist, explains why these components are qualified as being *independent.* Since edges of a digraph model information flows between interconnected systems, the fact that there exists no edge with tail outside an iSCC and head inside that iSCC means that the systems represented by the vertices within the iSCC are not influenced by the systems represented by vertices outside the iSCC. That is, the systems represented by the vertices contained in an iSCC form an *independent* subgroup within the network. In other words, those systems are not aware of systems outside the group. Note that any connected component of an undirected graph matches Definition 2.10 and therefore is an iSCC. Furthermore, it should be noted that Definition 2.10 does not exclude edges with tail inside an iSCC and head outside that iSCC.

It follows from the definition of an iSCC, that, at fixed time t, distinct iSCCs of digraphs are mutually independent, i.e., no paths exist connecting vertices contained in distinct iSCCs. However, there may exist a vertex not belonging to any iSCC that is a descendant of vertices belonging to distinct iSCCs. Such a situation occurs in the graph depicted in Figure 2.4: The iSCCs are the subgraphs induced by the vertex sets $\{v_2\}$ and $\{v_3, v_4\}$; vertex v_1 is a descendant of all other vertices. The graph depicted in Figure 2.1 has only one iSCC, namely the graph induced by the vertex set $\{v_1\}$. In the graph depicted in Figure 2.2, the unique iSCC coincides with the graph itself. Intuitively, the iSCCs identify the individuals which exert influence on other individuals in the network while themselves not being influenced by vertices outside the iSCC. In terms of consensus, the iSCCs identify the individuals to "control" in order to achieve agreement in the whole group. Below, we discuss important properties of iSCCs in digraphs eventually resulting in a generalization of Lemma 2.8 to digraphs. The following results are predominantly based on Wieland et al. (2008, 2010a).

It has been argued already, that – in contrast to undirected graphs where each vertex belongs to exactly one connected component – in digraphs there might be vertices which do not belong to any iSCC. An example is vertex v_1 in the graph depicted in Figure 2.4. However, iSCCs are well defined in the sense that any vertex belongs to at most one iSCC and any vertex not belonging to an iSCC is descendant of vertices that belong to an iSCC, as stated in the following theorems:

Theorem 2.11. *Let t be fixed time and $\mathcal{G}(t) = \{\mathcal{V}, \mathcal{E}(t), W(t)\}$ a digraph. Let $\tilde{\mathcal{G}}_i = \{\tilde{\mathcal{V}}_i, \tilde{\mathcal{E}}_i\}$, $i = 1, 2$ be two iSCCs of $\mathcal{G}(t)$.*

Then $\tilde{\mathcal{V}}_1 \bigcap \tilde{\mathcal{V}}_2 \neq \emptyset$ if and only if $\tilde{\mathcal{V}}_1 = \tilde{\mathcal{V}}_2$, i.e., $\tilde{\mathcal{G}}_1 = \tilde{\mathcal{G}}_2$.

Proof. The proof of sufficiency is immediate. To prove necessity, assume there exists a vertex $v \in \tilde{\mathcal{V}}_1 \bigcap \tilde{\mathcal{V}}_2$. Strong connectivity of iSCCs implies on the one hand that $\tilde{\mathcal{V}}_i \subseteq \{v\} \bigcup \mathcal{D}(v, \mathcal{G}(t))$, $i = 1, 2$ and on the other hand that $\mathcal{D}(v, \mathcal{G}(t)) \subseteq \mathcal{D}(w, \mathcal{G}(t))$ for any $w \in \tilde{\mathcal{V}}_1 \bigcup \tilde{\mathcal{V}}_2$. That is, all vertices of either of the two iSCCs are descendants of any vertex of either of the two iSCCs. By Definition 2.10 of an iSCC, or more specifically the property that iSCCs are maximal, subject to being strongly connected, this implies $\tilde{\mathcal{V}}_1 = \tilde{\mathcal{V}}_2$. □

Theorem 2.12. *Let t be fixed time and $\mathcal{G}(t) = \{\mathcal{V}, \mathcal{E}(t), W(t)\}$ a digraph. Let $v \in \mathcal{V}$.*

Then there exists at least one iSCC $\tilde{\mathcal{G}} = \{\tilde{\mathcal{V}}, \tilde{\mathcal{E}}\}$ such that either $v \in \tilde{\mathcal{V}}$ or $v \in \mathcal{D}(w, \mathcal{G}(t))$ for any $w \in \tilde{\mathcal{V}}$, denoted as $v \in \mathcal{D}(\tilde{\mathcal{V}}, \mathcal{G}(t))$.

Proof. The proof is given in Appendix A.3.1 □

Finally, we give the generalization of Lemma 2.8 to digraphs as follows.

Theorem 2.13. *Let t be fixed time and $\mathcal{G} = \{\mathcal{V}, \mathcal{E}(t), W(t)\}$ a digraph with $|\mathcal{V}| = N$ vertices, satisfying Assumption 2.2. Assume $\mathcal{G}(t)$ has $c \geq 1$ distinct iSCCs $\tilde{\mathcal{G}}_i = \{\tilde{\mathcal{V}}_i, \tilde{\mathcal{E}}_i\}$, $i \in \mathbb{N}_c$.*

Then, $\dim(\ker(L^T(t))) = c$ and there exist unique (modulo vertex permutations) vectors $p^i \in \mathbb{R}^N, i \in \mathbb{N}_c$ satisfying

$$\begin{aligned} p^i_j &> 0 \quad \text{if } v_j \in \tilde{\mathcal{V}}_i, \\ p^i_j &= 0 \quad \text{if } v_j \notin \tilde{\mathcal{V}}_i, \end{aligned} \qquad i \in \mathbb{N}_N \tag{2.2}$$

element-wise and $(p^i)^T 1_N = 1$, $i \in \mathbb{N}_c$, such that $\ker(L^T(t)) = \operatorname{span}(p^1, \ldots, p^c)$, i.e., the c vectors p^i, $i \in \mathbb{N}_c$ span a non-negative, orthogonal basis of the kernel of the transposed Laplacian matrix $L^T(t)$.

Proof. The proof is given in Appendix A.3.2. □

Theorem 2.13 conveys that the dimension of the kernel of the transposed Laplacian matrix $L^T(t)$ (and therefore also the Laplacian matrix $L(t)$) is given by the number of iSCCs of the corresponding digraph. This is very similar to the situation of undirected graphs, where the number of connected components determines the dimension of $\ker(L^T(t))$ (and $\ker(L(t))$) by Lemma 2.8. The kernel of $L^T(t)$ is spanned by a uniquely defined set of non-negative, orthogonal vectors p^i, $i \in \mathbb{N}_c$. In contrast to the case of undirected graphs, for which those vectors are specified in Lemma 2.8, Theorem 2.13 just states existence of those vectors with the aforementioned properties. The values taken by the elements of the vectors p^i, $i \in \mathbb{N}_c$ depend on the precise digraph $\mathcal{G}(t)$. The normalization conditions $(p^i)^T 1_N = 1$, $i \in \mathbb{N}_c$ are imposed to ensure uniqueness of the vectors p^i, $i \in \mathbb{N}_c$. Since the vectors p^i, $i \in \mathbb{N}_c$ are unique, a possibility to determine the iSCCs of a given graph $\mathcal{G}(t)$ is to construct a non-negative, orthogonal basis of the kernel of $L^T(t)$. Then (2.2) determines the vertex sets of the iSCCs of $\mathcal{G}(t)$.

In analogy to Corollary 2.9 for undirected graphs, Theorem 2.13 has an immediate corollary relating the spectrum of $L(t)$ to graph connectivity.

Corollary 2.14. *Let t be fixed time and $\mathcal{G} = \{\mathcal{V}, \mathcal{E}(t), W(t)\}$ a digraph satisfying Assumption 2.2.*

Then $\mathcal{G}(t)$ is connected at time t if and only if the second eigenvalue $\lambda_2(L(t))$ of the Laplacian $L(t)$ has positive real part, i.e., $\mathfrak{Re}(\lambda_2(L(t))) > 0$, where the second eigenvalue $\lambda_2(L(t))$ is the eigenvalue of $L(t)$ with the second smallest real part by convention.

Proof. ($\Rightarrow$) If $\mathcal{G}(t)$ is connected, it has at most one iSCC; by Theorem 2.12, it must have at least one iSCC, thus it has exactly one iSCC. Then, by Theorem 2.13, 0 is a simple eigenvalue of $L(t)$ and thus all other eigenvalues, and in particular the second eigenvalue, of $L(t)$ have positive real part.

($\Leftarrow$) If $\mathfrak{Re}(\lambda_2(L(t))) > 0$, 0 is a simple eigenvalue of $L(t)$ which implies that the graph has exactly one iSCC by Theorem 2.13. By Theorem 2.12, any vertex contained in that iSCC must be a centroid of $\mathcal{G}(t)$, i.e., the graph is connected. □

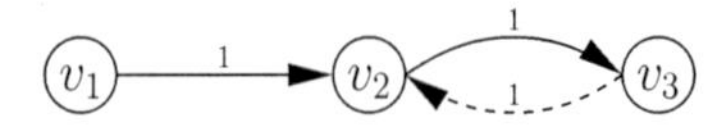

Figure 2.5: Digraph where an additional edge decreases the second smallest real part $\mathfrak{Re}(\lambda_2(L(t)))$ of the Laplacian spectrum. Without the dashed edge, $\mathfrak{Re}(\lambda_2(L(t))) = 1$, with the dashed edge, $\mathfrak{Re}(\lambda_2(L(t))) = \frac{3-\sqrt{5}}{2} < 1$.

Corollary 2.14 suggests that the second smallest real part $\mathfrak{Re}(\lambda_2(L(t)))$ of the Laplacian $L(t)$ of a digraph $\mathcal{G}(t)$ plays, to some extent, the role of the algebraic connectivity of undirected graphs. Most importantly, we know that this quantity is positive if and only if the corresponding graph is strongly connected and zero otherwise. It should be noted that, for undirected graphs, $\mathfrak{Re}(\lambda_2(L(t))) = \lambda_2(L(t))$ because $L(t) = L^T(t)$ and thus $\sigma(L(t)) \subset \mathbb{R}$. That is the second smallest real part $\mathfrak{Re}(\lambda_2(L(t)))$ of the Laplacian $L(t)$ is indeed a generalization of the algebraic connectivity to digraphs. However, the value $\mathfrak{Re}(\lambda_2(L(t)))$ does not possess many of the nice properties of the algebraic connectivity for undirected graphs, e.g., $\mathfrak{Re}(\lambda_2(L(t)))$ may decrease when edges are added to a graph as in the graph depicted in Figure 2.5. An alternative proof of Corollary 2.14 is given in Ren and Beard (2005).

Other generalizations of the algebraic connectivity to directed graphs exist in the literature. For completeness, we mention here the definition due to Wu (2005), who defines the algebraic connectivity with the help of a Rayleigh-Ritz ratio (see Horn and Johnson, 1985) as

$$a(t) = \min_{\substack{x \in \mathbb{R}^N \\ x \neq 0 \\ x \perp 1_N}} \frac{x^T L(t) x}{x^T x}.$$

For undirected graphs, $a(t) = \lambda_2(L(t))$, i.e., the quantity $a(t)$ is indeed a generalization of the algebraic connectivity to digraphs. This definition of algebraic connectivity is used for consensus and synchronization problems in Scardovi et al. (2010). However, for general digraphs, no simple relation exists between the algebraic connectivity $a(t)$ and graph connectedness according to Definitions 2.4 and 2.5. Therefore, the algebraic connectivity $a(t)$ is not used for the results in this thesis.

2.3 Summary

In this chapter we introduced basic concepts from graph theory and more specifically algebraic graph theory as far as they are relevant for consensus and synchronization problems.

We started by introducing the most important definitions, including several definitions for graph connectedness in fixed and time-varying communication topologies. Furthermore, a link between the graph Laplacian matrix and the Vicsek model was highlighted.

The second part of this chapter was devoted to generalizations of concepts from algebraic graph theory for undirected graphs to digraphs as they will be used to model communication topologies later on (see Wieland et al., 2010a). Particularly, we developed the concept of iSCCs of digraphs as a generalization to components of undirected graphs.

Chapter 3

Consensus with Static Diffusive Couplings

In the previous chapter, a close link between linear diffusive couplings and algebraic graph theory, more specifically the Laplacian matrix $L(t)$ of the communication graph $\mathcal{G}(t) = \{\mathcal{V}, \mathcal{E}(t), W(t)\}$, has been established. In particular, we argued that the dynamics of the Vicsek model is essentially characterized by the graph Laplacian $L(t)$. In this chapter, we will have a closer look on the dynamics for consensus models inspired by the Vicsek model and different generalizations. Particularly, we will quantify convergence properties of the consensus dynamics.

Static diffusive couplings have been used in several setups. For instance, Slotine and Wang (2004), Stan and Sepulchre (2007), Wang and Slotine (2005) use static diffusive couplings to synchronize nonlinear oscillators over constant, balanced communication graphs. Arcak (2007), Chopra and Spong (2006), Pogromsky (1997) use diffusive couplings to synchronize passive systems. Nijmeijer and Mareels (1997) relate the observer design problem to master-slave synchronization with diffusive couplings. Another example is of course the Kuramoto model (Kuramoto, 1975) introduced in Section 1.2. The methods presented in this chapter are restricted to individuals modeled as LTI systems.

Inspired by the results obtained for the Vicsek model (1.1), we start with a continuous time consensus model given as

$$\dot{x}(t) = u(t) \tag{3.1a}$$

$$u(t) = -L(t)x(t), \tag{3.1b}$$

where $x(t) = (x_1(t), \ldots, x_N(t))^T \in \mathbb{R}^N$ is the stacked vector of the scalar states $x_k(t) \in \mathbb{R}$, $k \in \mathbb{N}_N$ of the individual systems, $u(t) = (u_1(t), \ldots, u_N(t))^T \in \mathbb{R}^N$ is the stacked vector of the inputs $u_k(t) \in \mathbb{R}$, $k \in \mathbb{N}_N$ to the individual systems, and $L(t) \in \mathbb{R}^{N \times N}$ is the time-varying Laplacian matrix of some communication graph $\mathcal{G}(t) = \{\mathcal{V}, \mathcal{E}(t), W(t)\}$ with $|\mathcal{V}| = N$ (see Jadbabaie et al., 2003). The above model (3.1) consists of N open loop integrator systems (3.1a) and diffusive couplings realized as a static state feedback (3.1b). It can be viewed as the prototype for many consensus models. The feedback is determined by the communication graph $\mathcal{G}(t)$ represented by its Laplacian matrix $L(t)$.

Consensus in the above model (3.1) means that all individual systems asymptotically reach a common value. That is consensus is asymptotically achieved if $(x_k(t) - x_j(t)) \to 0$ as $t \to \infty$ for all $j, k \in \mathbb{N}_N$ or equivalently if $(x(t) - 1_N\xi(t)) \to 0$ as $t \to \infty$ for some real valued function $\xi(\cdot) : \mathbb{R} \to \mathbb{R}$. If the individual systems converge to consensus, this implies that $u(t) \to 0$ as $t \to \infty$, thus one expects that $x(t)$ asymptotically reaches a constant vector and $\xi(t)$ can be chosen constant, i.e., $\xi(t) \equiv \xi_0$ for some $\xi_0 \in \mathbb{R}$ in case of the model described by (3.1).

An important generalization of the prototype model (3.1) consists in considering individual systems whose models are given by more general dynamical systems. In this chapter, we thus consider networks of N identical individuals, modeled as general LTI systems

$$\dot{x}_k(t) = Ax_k(t) + Bu_k(t), \quad k \in \mathbb{N}_N, \tag{3.2}$$

with state vectors $x_k(t) \in \mathbb{R}^n$ and input vector $u_k(t) \in \mathbb{R}^p$. We are interested in the effect induced by static, diffusive couplings

$$u_k(t) = K \sum_{j=1}^{N} w_{k,j}(t)(x_k(t) - x_j(t)), \quad k \in \mathbb{N}_N \tag{3.3}$$

on the network of systems (3.2). In the couplings (3.3), the matrix $K \in \mathbb{R}^{p \times n}$ is a feedback gain, which we consider as a degree of freedom in the design methods proposed in this chapter; the values $w_{k,j}(t) \in \overline{\mathbb{R}}_+$, $j, k \in \mathbb{N}_N$ are the elements of the adjacency matrix $W(t) = [w_{k,j}(t)] \in \mathbb{R}^{N \times N}$ of some communication graph $\mathcal{G}(t) = \{\mathcal{V}, \mathcal{E}(t), W(t)\}$ with $|\mathcal{V}| = N$. We would like to stress that the couplings defined by (3.3) are indeed diffusive couplings as defined in Section 1.2 because they are proportional to a weighted (by the coefficients $w_{k,j}(t)/\sum_{j=1}^{N} w_{k,j}(t)$, $j, k \in \mathbb{N}_N$) average of relative states $x_k(t) - x_j(t)$, $j, k \in \mathbb{N}_N$.

Similarly to (3.1), we frequently consider the ensemble of the states $x_k(t) \in \mathbb{R}^n$, $k \in \mathbb{N}_N$ of the individual systems as a single, large state vector and write the whole network as one big system. Using the Kronecker product[1], a compact notation of the network of systems (3.2) with couplings (3.3) is obtained as

$$\dot{x}(t) = (I \otimes A)x(t) + (I \otimes B)u(t) \tag{3.4a}$$

$$u(t) = (L(t) \otimes K)x(t) \tag{3.4b}$$

with stacked state vector $x(t) = (x_1^T(t), \dots, x_N^T(t))^T \in \mathbb{R}^{Nn}$ and stacked input vector $u(t) = (u_1^T(t), \dots, u_N^T(t))^T \in \mathbb{R}^{Np}$ (see Fax and Murray, 2004, Wieland et al., 2010a). The system description (3.4) possesses a useful property. Namely, despite the fact that (3.4) is a global representation of the network, the local coupling structure and the system dynamics of the individual network members stay manifest in the system equations.

The model (3.4) includes several important special cases, some of which we would like to mention below. A first example is the prototype model (3.1), which is obtained from (3.4) by setting $n = p = 1$, $A = 0$, $B = 1$ and $K = -1$.

A second important case is consensus among chains of integrators, which was treated, e.g., in Hu and Lin (2010), Ren and Atkins (2005), Ren et al. (2006), Yang and Fang (2010). This particular case is obtained from (3.4) by setting $p = 1$ and

$$A = \begin{pmatrix} 0 & 1 & & 0 \\ & \ddots & \ddots & \\ & & 0 & 1 \\ 0 & & & 0 \end{pmatrix}, \quad B = \begin{pmatrix} 0 \\ \vdots \\ 0 \\ 1 \end{pmatrix}.$$

[1] Details on the Kronecker product are found in Horn and Johnson (1991). We will frequently use the identiy $(A \otimes B)(C \otimes D) = (AC) \otimes (BD)$ for matrices $A \in \mathbb{R}^{m \times n}$, $B \in \mathbb{R}^{p \times q}$, $C \in \mathbb{R}^{n \times k}$, and $D \in \mathbb{R}^{q \times r}$.

The interest in consensus of individual systems modeled by chains of integrators is motivated, e.g., by the observation that flocking behavior may require consensus about positions, speed, and acceleration among the individuals (see Ren et al., 2006).

A third example which received considerable attention in literature (Ren, 2008, Scardovi and Sepulchre, 2009) is consensus (or synchronization) of harmonic oscillators. This corresponds to (3.4a) with $n = 2$ and

$$A = \begin{pmatrix} 0 & \omega \\ -\omega & 0 \end{pmatrix}$$

for some frequency of oscillation $\omega \in \mathbb{R} \setminus \{0\}$.

Consensus with general LTI systems (3.2) thus represents a unifying picture for several relevant special cases and thereby contributes to a deepened understanding of linear consensus problems.

3.1 Definition of Consensus and Problem Setup

As before, we define consensus as the property that $(x_k(t) - x_j(t)) \to 0$ as $t \to \infty$ for all $j, k \in \mathbb{N}_N$. More formally, state consensus can be defined as follows:

Definition 3.1 (State Consensus). *The N systems* (3.2) *are said to* asymptotically reach state consensus, *if there exists some trajectory $\xi(\cdot) : \mathbb{R} \to \mathbb{R}^n$ such that*

$$\lim_{t \to \infty} \|x_k(t) - \xi(t)\| = 0 \tag{3.5}$$

for all $k \in \mathbb{N}_N$. In that case, the trajectory $\xi(\cdot)$ is called a consensus trajectory.

The consensus trajectory $\xi(\cdot)$ in the above definition is a curve in $\mathbb{R}^n$ parameterized by time t. Condition (3.5) expresses that consensus is reached if the state trajectories $x_k(\cdot)$, $k \in \mathbb{N}_N$ of all individual systems asymptotically converge to the common consensus trajectory. However, Definition 3.1 makes no statement about properties of that consensus trajectory besides being close to state trajectories $x_k(t)$ for $t \to \infty$. In particular, the consensus trajectory is not directly related to the dynamics of the individual systems in the network. Definition 3.1 does not even require the consensus trajectory to be a solution of any linear ODE system. Since for a given communication graph $\mathcal{G}(t)$ the solutions $x_k(t)$, $k \in \mathbb{N}_N$ are determined by initial conditions $x_k(0)$, $k \in \mathbb{N}_N$, it is natural that the consensus trajectory $\xi(t)$ will also depend on initial conditions $x_k(0)$, $k \in \mathbb{N}_N$. However, for fixed initial conditions $x_k(0) \in \mathbb{R}^n$, $k \in \mathbb{N}_N$ consensus trajectories are not unique. This fact is readily verified as follows: let $\xi(t)$ be a consensus trajectory for some initial conditions $x_k(0) \in \mathbb{R}^n$, $k \in \mathbb{N}_N$ and let $\psi(\cdot) : \mathbb{R} \to \mathbb{R}^n$ be any mapping such that $\lim_{t\to\infty} \psi(t) = 0$, then $\tilde{\xi}(t) \triangleq \xi(t) + \psi(t)$ is also a consensus trajectory. Yet, assuming an exponential convergence rate in (3.5), the specific couplings used in (3.4) imply that, if there exists any consensus trajectory, there exists a particular, canonical consensus trajectory. This fact is formally stated in the following lemma:

Lemma 3.2. *Let $\tilde{\xi}(\cdot) : \mathbb{R} \to \mathbb{R}^n$ be a consensus trajectory for* (3.4) *for fixed initial conditions $x_k(0) \in \mathbb{R}^n$, $k \in \mathbb{N}_N$.*

If (3.5) *is satisfied with an exponential convergence rate, i.e., there exist constants* $M_1, \mu_1 \in \mathbb{R}_+$ *such that*

$$\|x_k(t) - \tilde{\xi}(t)\| \leq M_1 e^{-\mu_1 t} \max_{j \in \mathbb{N}_N} \|x_j(0) - \tilde{\xi}(0)\|$$

for all $k \in \mathbb{N}_N$*, then there exists* $\xi_0 \in \mathbb{R}^n$ *and constants* $M_2, \mu_2 \in \mathbb{R}_+$ *such that*

$$\|x_k(t) - e^{At}\xi_0\| \leq M_2 e^{-\mu_2 t} \max_{j \in \mathbb{N}_N} \|x_j(0) - \xi_0\|$$

for all $k \in \mathbb{N}_N$*, i.e., the solution* $\xi(t) = e^{At}\xi_0$ *of the open-loop dynamics* (3.2) *with* $u_k(t) \equiv 0$ *is a consensus trajectory for* (3.4) *with exponential convergence rate for the same initial conditions* $x_k(0) \in \mathbb{R}^n$, $k \in \mathbb{N}_N$.

Proof. If $\tilde{\xi}(\cdot)$ is a consensus trajectory, then by condition (3.5)

$$\lim_{t \to \infty} \|x(t) - (1_N \otimes \tilde{\xi}(t))\| = 0,$$

i.e., $x(t)$ asymptotically converges to the subspace $\mathrm{im}(1_N \otimes I_n) \subset \mathbb{R}^{Nn}$. If $x(t) \in \mathrm{im}(1_N \otimes I_n)$, then $x(t) = 1_N \otimes \xi(t)$ for some trajectory $\xi(\cdot) : \mathbb{R} \to \mathbb{R}^n$. We obtain $u(t) = (L(t)1_N) \otimes (K\xi(t)) = 0$ and $\dot{x}(t) = 1_N \otimes \dot{\xi}(t) = 1_N \otimes (A\xi(t)) \in \mathrm{im}(1_N \otimes I_n)$. That is, $\mathrm{im}(1_N \otimes I_n)$ is an invariant subspace of dimension n with dynamics described by the open loop dynamics of the individual system, i.e., (3.2) with $u_k(t) \equiv 0$.

Convergence of $x(t)$ to the subspace $\mathrm{im}(1_N \otimes I_n) \subset \mathbb{R}^{Nn}$ is exponential by assumption. By Lemma B.1 given in Appendix B, this implies that $x(t)$ exponentially converges to a particular solution $1_N \otimes \xi(t)$ contained in $\mathrm{im}(1_N \otimes I_n)$. This completes the proof. □

The key property of the diffusive couplings (3.4b) used in the proof of Lemma 3.2 is that the couplings vanish once consensus is reached. This results from the fact that diffusive couplings exclusively depend on relative information. Lemma 3.2 expresses that, under the assumption that consensus is reached exponentially fast, the synchronized solutions of all individual systems behave like the individual system without input. This property is exactly what one would intuitively expect to happen: Consensus algorithm (3.3) does not alter the system behavior but just 'aligns' the individual solutions in a network of systems.

In the context of static, diffusive couplings (3.4b), there are two problems of potential interest that are related to consensus in the sense of Definition 3.1 and that we will be considering subsequently. The first problem is an analysis problem and can be posed as follows:

Problem 3.3 (State Consensus Analysis). *Let the* N *individual systems* (3.2) *be modeled by given matrices* $A \in \mathbb{R}^{n \times n}$ *and* $B \in \mathbb{R}^{n \times p}$ *and let the diffusive couplings* (3.3) *be defined by some given coupling gain* $K \in \mathbb{R}^{p \times n}$ *and some communication graph* $\mathcal{G}(t) = \{\mathcal{V}, \mathcal{E}(t), W(t)\}$ *with* $|\mathcal{V}| = N$ *belonging to a given class.*

Decide whether state consensus is asymptotically reached for all initial conditions $x_k(0)$, $k \in \mathbb{N}_N$ *of the individual systems and, in case the answer is yes, what the consensus trajectory* $\xi(\cdot)$ *will be.*

If consensus is reached exponentially fast, the question about the consensus trajectory $\xi(\cdot)$ is partly answered by Lemma 3.2. But in some instances, it is possible to specify the initial condition $\xi_0 = \xi(0)$ of the consensus trajectory in dependence of the initial conditions $x_k(0)$, $k \in \mathbb{N}_N$ of the individual systems and the communication graph $\mathcal{G}$. A slight modification of the State Consensus Analysis Problem 3.3 consists in considering the question under which conditions on the communication graph $\mathcal{G}(t)$ consensus is reached.

The second problem is the corresponding design problem given below.

Problem 3.4 (State Consensus Design). *Let the N individual systems* (3.2) *be modeled by given matrices $A \in \mathbb{R}^{n\times n}$ and $B \in \mathbb{R}^{n\times p}$. Let the communication topology in the diffusive couplings* (3.3) *be defined by some communication graph $\mathcal{G}(t) = \{\mathcal{V}, \mathcal{E}(t), W(t)\}$ with $|\mathcal{V}| = N$ belonging to a given class.*

Find, if possible, a coupling gain $K \in \mathbb{R}^{p\times n}$ such that consensus is asymptotically reached for all initial conditions $x_k(0)$, $k \in \mathbb{N}_N$ of the individual systems.

In the following sections, we will propose different solutions for Problems 3.3 and 3.4, first for *fixed communication topologies*, i.e., constant communication graphs and thus constant graph Laplacian matrices in Section 3.2, and subsequently for *time-varying communication topologies* in Section 3.3. In Section 3.4, we will comment on the problem of *consensus with a leader* and, in this context, consider a slightly different version of the State Consensus Design Problem 3.4, where we also consider parts of the communication graph as degree of freedom in the design. To conclude this chapter, we will shortly summarize the results in Section 3.5.

3.2 Consensus for Fixed Communication Topologies

In this section, we consider the problem of consensus under fixed communication topologies, i.e., constant communication graphs. We therefore omit the parameter t in the description of the communication graph and simply write $\mathcal{G} = \{\mathcal{V}, \mathcal{E}, W\}$. Throughout this section, we assume without loss of generality that the communication graph $\mathcal{G}$ satisfies Assumption 2.2, i.e., $(v_j, v_k) \in \mathcal{E}$ if and only if $w_{k,j} > 0$ for all $j, k \in \mathbb{N}_{|\mathcal{V}|}$.

3.2.1 Integrator Consensus for Fixed Communication Topologies

We start by discussing a solution to Problem 3.3 for the prototype model (3.1). The interest in starting with this special case lies primarily in its simplicity. Insight gained from the special case of this simple model will be helpful when generalizing to the model (3.4) later on. Our first solution to the State Consensus Analysis Problem 3.3 is given in the following lemma:

Lemma 3.5. *Let $\mathcal{G} = \{\mathcal{V}, \mathcal{E}, W\}$ be a constant communication graph with $|\mathcal{V}| = N > 1$ and Laplacian matrix L satisfying Assumption 2.2.*

Then consensus in the prototype model (3.1) *is reached for all initial conditions $x(0) \in \mathbb{R}^N$ if and only if the graph $\mathcal{G}$ is connected.*

In that case, there exists a unique, element-wise non-negative vector $p \in \mathbb{R}^N$ satisfying $p^T 1_N = 1$ and $p^T L^T = 0$ and constants $M, \mu \in \mathbb{R}_+$, such that

$$\|x(t) - 1_N p^T x(0)\| \leq M e^{-\mu t} \|x(0) - 1_N p^T x(0)\| \tag{3.6}$$

i.e., the consensus value is given by $\xi_0 = p^T x(0)$*, and it is reached exponentially fast.*

Proof. The set where consensus is reached is the subspace $\mathrm{im}(1_N) \subset \mathbb{R}^N$. By the properties of the Laplacian matrix L, namely $L1_N = 0$, this subspace is invariant for (3.1). Consensus is reached for all initial conditions if and only if all modes of (3.1) not contained in $\mathrm{im}(1_N)$ are asymptotically stable which is the case if and only if 0 is a simple eigenvalue of L and all other eigenvalues of L have positive real part. By Corollary 2.14 this is equivalent to $\mathcal{G}$ being connected.

If 0 is a simple eigenvalue of L, there exists a unique element-wise non-negative vector $p \in \mathbb{R}^N$ such that $p^T 1_N = 1$ and $p^T L = 0$ by Theorem 2.13, i.e., p^T is the normalized left-eigenvector of L for the eigenvalue 0. Thus, it defines the projection of the initial condition $x(0) \in \mathbb{R}^N$ to the mode contained in the consensus subspace $\mathrm{im}(1_N)$. Since all other modes are asymptotically stable and thus exponentially stable, the inequality (3.6) follows. □

In simple words, Lemma 3.5 states the well-known result that connectedness of the constant communication graph $\mathcal{G}$ is necessary and sufficient for consensus in the prototype model (3.1). In view of Theorem 2.13, this means that consensus is achieved if and only if the graph $\mathcal{G}$ has exactly one iSCC. In that case, all states $x_k(t)$ converge to a common constant value $\xi_0 = p^T x(0)$, where $p \in \mathbb{R}^N$ is the unique vector satisfying $p^T L = 0$ and $p^T 1_N = 1$. Again by Theorem 2.13, we know that $p_k \geq 0$ with $p_k > 0$ if and only if the corresponding vertex $v_k \subset \mathcal{V}$ belongs to the unique iSCC of $\mathcal{G}$ for $k \in \mathbb{N}_N$. That is, the information initially available at the members of the unique iSCC determines the ultimate value all individuals will eventually converge to.

To understand this observation, remember that an iSCC is a strongly connected induced subgraph $\tilde{\mathcal{G}}$ with some subset $\tilde{\mathcal{V}} \subset \mathcal{V}$ of vertices such that no path exists in the complete graph $\mathcal{G}$ starting at some vertex $v \in \mathcal{V} \setminus \tilde{\mathcal{V}}$ outside the iSCC and ending at some vertex $w \in \tilde{\mathcal{V}}$ within the iSCC. That is information from systems that are represented by vertices not belonging to an iSCC cannot be transmitted through the communication graph $\mathcal{G}$ to systems represented by vertices belonging to the iSCC. Therefore, if all systems eventually agree upon some value, this value cannot depend on information that was not initially available to some system represented by a vertex belonging to the unique iSCC of $\mathcal{G}$. This fact is reflected by $p_k = 0$ if the corresponding vertex $v_k \in \mathcal{V}$ does not belong to the iSCC (cf. Theorem 2.13). On the other hand, within the iSCC, paths exist between any pair of distinct vertices, i.e., information can be exchanged between any pair of distinct vertices. Thus, all systems represented by vertices belonging to the iSCC contribute to the final value, the group agrees upon. This contribution is quantified by the corresponding elements $p_k > 0$ of the vector p.

A particular case of consensus in the prototype model (3.1) is *average consensus*, i.e., the states $x_k(t)$, $k \in \mathbb{N}_N$ converge to the arithmetic mean of the initial conditions $x_k(0), k \in \mathbb{N}_N$, formally

$$\lim_{t\to\infty} \left\| x(t) - 1_N \frac{1_N^T x(0)}{N} \right\| = 0.$$

Necessary and sufficient conditions for average consensus are given in the corollary below.

Corollary 3.6. *Let* $\mathcal{G} = \{\mathcal{V}, \mathcal{E}, W\}$ *be a constant communication graph with* $|\mathcal{V}| = N > 1$ *and Laplacian matrix* L *satisfying Assumption 2.2.*

Then average consensus in the prototype model (3.1) *is reached for all initial conditions* $x(0) \in \mathbb{R}^N$ *if and only if the graph* $\mathcal{G}$ *is balanced and connected.*

Proof. By Lemma 3.5, average consensus is reached if and only if the communication graph is connected and $p = 1_N$ satisfies $p^T L = 0$, i.e., 1_N^T is a left-eigenvector of the Laplacian matrix L corresponding to the simple eigenvalue 0. By definition of the Laplacian matrix,

$$1_N^T L = \left(\sum_{j=1}^N w_{1,j} - \sum_{j=1}^N w_{j,1}, \ldots, \sum_{j=1}^N w_{N,j} - \sum_{j=1}^N w_{j,N} \right).$$

Thus, $1_N^T L = 0$ is equivalent to $\mathcal{G}$ being balanced. □

With these first solutions for Problem 3.3 at hand, we now proceed with solutions for Problems 3.3 and 3.4 in the case of individuals modeled as general, identical LTI systems.

3.2.2 Consensus among Identical Linear Systems with Fixed Communication Topologies

The objective of this section is to generalize and extend the results given in the previous section for the prototype model (3.1) to networks of identical LTI systems (3.4). We start by a solution to the State Consensus Problem 3.3 generalizing Lemma 3.5. The result below is essentially due to Fax and Murray (2004). It has also been reported independently in Wieland and Allgöwer (2010), Wieland et al. (2008) and partly in Tuna (2008a).

Theorem 3.7. *Let* $\mathcal{G} = \{\mathcal{V}, \mathcal{E}, W\}$ *be a constant communication graph with* $|\mathcal{V}| = N > 1$ *and Laplacian matrix* L *satisfying Assumption 2.2. Let* $A \in \mathbb{R}^{n \times n}$, $B \in \mathbb{R}^{n \times p}$, *and* $K \in \mathbb{R}^{p \times n}$.

Then the individual systems in the model (3.4) *reach consensus for all initial conditions* $x_k(0) \in \mathbb{R}^n$, $k \in \mathbb{N}_N$ *if and only if the systems*

$$\dot{x}(t) = (A + \lambda_k(L)BK)x(t), \quad k \in \mathbb{N}_N \setminus \{1\} \tag{3.7}$$

with state vector $x(t) \in \mathbb{R}^n$ *are asymptotically stable.*

In that case, there exists an element-wise non-negative vector $p \in \mathbb{R}^N$ *satisfying* $p^T 1_N = 1$ *and* $p^T L^T = 0$ *and constants* $M, \mu \in \mathbb{R}_+$, *such that*

$$\left\| x(t) - 1_N \otimes \left[e^{At} \left((p^T \otimes I_n) x(0) \right) \right] \right\| \leq M e^{-\mu t} \left\| x(0) - 1_N \otimes \left((p^T \otimes I_n) x(0) \right) \right\| \tag{3.8}$$

i.e., $\xi(t) = e^{At} \left((p^T \otimes I_n) x(0) \right)$ *is a consensus trajectory and convergence to consensus is exponentially fast.*

Proof. The set where consensus is reached is the subspace $\mathrm{im}(1_N \otimes I_n) \subset \mathbb{R}^{Nn}$. As in the proof for Lemma 3.5, the properties of the Laplacian matrix imply that this subspace is invariant for (3.4). In fact

$$(L \otimes K)(1_N \otimes I_n) = (L 1_N \otimes K) = 0,$$

i.e., for $x(t) \in \mathrm{im}(1_N \otimes I_n)$, the couplings vanish and $u(t) \equiv 0$. Furthermore, this implies that for $x(t) \in \mathrm{im}(1_N \otimes I_n)$, the dynamic evolution is governed by $\dot{x}(t) = (I_N \otimes A)x(t)$, i.e., all systems are decoupled and evolve according to their open loop dynamics.

The projection of the initial conditions $x(0)$ to the modes contained in $\mathrm{im}(1_N \otimes I_n)$ is defined by some matrix $(p^T \otimes I_n)$ for some vector $p \in \mathbb{R}^N$ satisfying $p^T 1_N = 1$ and $p^T L = 0$. By Theorem 2.13, we can choose p to be element-wise non-negative. This implies that, if consensus is reached exponentially, then inequality (3.8) is satisfied.

Let $T \in \mathbb{C}^{N\times N}$ be a transformation matrix such that $\Lambda \triangleq T^{-1}LT$ is upper triangular and the elements on the diagonal of Λ are the eigenvalues $\lambda_k(L)$, $k \in \mathbb{N}_N$ ordered by increasing real parts. We can choose T such that

$$T\begin{pmatrix}1\\0\\\vdots\\0\end{pmatrix} = 1_N, \quad (T^{-1})^T\begin{pmatrix}1\\0\\\vdots\\0\end{pmatrix} = p,$$

i.e., the first column of T is the all ones vector and the first row of T^{-1} equals p^T.

Consider the change of coordinates $x(t) = (T \otimes I_n)z(t)$. Then

$$\dot{z}(t) = ((I_N \otimes A) + (\Lambda \otimes BK))z(t),$$

where $(I_N \otimes A) + (\Lambda \otimes BK)$ is an upper block-triangular matrix. Due to the particular choice of T, the first $(n \times n)$ diagonal block of $(I_N \otimes A) + (\Lambda \otimes BK)$ corresponds to the dynamics of (3.4) restricted to the consensus subspace $\mathrm{im}(1_N \otimes I_n)$.

Consensus is asymptotically (or equivalently exponentially) reached for all initial conditions $x(0) \in \mathbb{R}^{Nn}$ if and only if all modes of (3.4) not contained in $\mathrm{im}(1_N \otimes I_n)$ are asymptotically stable. The stability of these modes is determined by the stability of the remaining diagonal blocks which is exactly the stability of the systems (3.7). □

Theorem 3.7 establishes equivalence between consensus in the network described by (3.4) and asymptotic stability of $(N-1)$ LTI systems of order n (3.7). Those systems are obtained from the description of the members (3.2) of the network with state feedback $u(t) = \lambda_k(L)Kx(t)$, $k \in \mathbb{N}_N\backslash\{1\}$, i.e., all $N-1$ systems are equivalent except for different gains in the feedback. Those gains are determined by the eigenvalues of the Laplacian matrix L encoding the communication topology.

The gains $\lambda_k(L)$, $k \in \mathbb{N}_N\backslash\{1\}$ may take complex values. Therefore, a remark is in order concerning stability of the systems (3.7). Just as in case of LTI systems with real dynamics matrices, the systems (3.7) are asymptotically stable if and only if all eigenvalues of the complex matrices $A + \lambda_k(L)BK \in \mathbb{C}^{n\times n}$ have negative real part. Instead of checking stability of these complex matrices, an equivalent condition involving only real quantities is readily obtained by observing that $\sigma(A + \lambda_k(L)BK) \subset \mathbb{C}_-$ if and only if the real $(2n \times 2n)$ matrix

$$\begin{pmatrix} A + \mathfrak{Re}(\lambda_k(L))BK & \mathfrak{Im}(\lambda_k(L))BK \\ -\mathfrak{Im}(\lambda_k(L))BK & A + \mathfrak{Re}(\lambda_k(L))BK \end{pmatrix}$$

is Hurwitz. The latter condition can be checked with standard stability tests for linear systems.

Theorem 3.7 does not relate consensus to graph connectedness directly. However, in Corollary 2.14, a relation between the eigenvalues of the Laplacian matrix and connectedness was established. Namely, 0 is a simple eigenvalue of L if and only if the corresponding graph is connected. We use this observation to relate graph connectedness to consensus in LTI systems.

Corollary 3.8. *Let $\mathcal{G} = \{\mathcal{V}, \mathcal{E}, W\}$ be a constant communication graph with $|\mathcal{V}| = N > 1$ and Laplacian matrix L satisfying Assumption 2.2. Let $A \in \mathbb{R}^{n \times n}$, $B \in \mathbb{R}^{n \times p}$, and $K \in \mathbb{R}^{p \times n}$.*

If the individual systems in the model (3.4) *reach consensus for all initial conditions $x_k(0) \in \mathbb{R}^n$, $k \in \mathbb{N}_N$, then the pair (A, B) is stabilizable and the communication graph $\mathcal{G}$ is connected or the matrix A is Hurwitz.*

Proof. Stability of the systems (3.7) implies stabilizability of the pair (A, B).

It remains to show that consensus implies either connectedness of the communication graph $\mathcal{G}$ or the matrix A being Hurwitz. Assume the graph $\mathcal{G}$ is not connected and A is not Hurwitz. Then, by Corollary 2.14, $\lambda_2(L) = 0$, thus $\dot{x} = A + \lambda_2(L)BKx = Ax$ is not asymptotically stable. Therefore, by Theorem 3.7, consensus is not reached for all initial conditions. □

In Corollary 3.8, we established a necessary condition for the model (3.4) reaching consensus for all initial conditions $x(0) \in \mathbb{R}^{Nn}$. Namely, consensus can only be reached if either A is Hurwitz, i.e., the individual systems (3.2) are asymptotically stable or the communication topology in the network is represented by a connected communication graph and (A, B) is stabilizable, i.e., the individual systems (3.2) are stabilizable.

We want to shortly comment on the case when A is Hurwitz and $\mathcal{G}$ is not necessarily connected. If consensus is reached, we know by Theorem 3.7, that $\xi(t) = e^{At}(p^T \otimes I_n)x(0)$ is a consensus trajectory. Furthermore, since A is Hurwitz, there exist constants $K, \mu \in \mathbb{R}_+$ such that $\|\xi(t)\| \leq Ke^{-\mu t}\|\xi(0)\|$. This implies that $\tilde{\xi}(t) \equiv 0$ is a consensus trajectory as well, which does not depend on initial conditions $x(0) \in \mathbb{R}^{Nn}$. That is, all individual systems asymptotically converge to the origin or, in other words, the network (3.4) as a whole is asymptotically stable. Such a type of behavior, is easily obtained without any couplings. It is thus considered the trivial case in the context of consensus problems.

Definition 3.9 (Non-Trivial Consensus). *Let $\mathcal{G} = \{\mathcal{V}, \mathcal{E}, W\}$ be a constant communication graph with $|\mathcal{V}| = N > 1$ and Laplacian matrix L. Let $A \in \mathbb{R}^{n \times n}$, $B \in \mathbb{R}^{n \times p}$, and $K \in \mathbb{R}^{p \times n}$.*

The network (3.4) *is said to reach* non-trivial consensus *if it reaches consensus for all initial conditions $x(0) \in \mathbb{R}^{Nn}$ and there exists some initial condition $x(0) \in \mathbb{R}^{Nn}$ such that $\xi(t) \equiv 0$ is not a consensus trajectory for* (3.4).

With the above definition, we can refine Corollary 3.8 as follows:

Corollary 3.10. *Let $\mathcal{G} = \{\mathcal{V}, \mathcal{E}, W\}$ be a constant communication graph with $|\mathcal{V}| = N > 1$ and Laplacian matrix L satisfying Assumption 2.2. Let $A \in \mathbb{R}^{n \times n}$, $B \in \mathbb{R}^{n \times p}$, and $K \in \mathbb{R}^{p \times n}$.*

If the individual systems in the model (3.4) *reach non-trivial consensus, then the pair (A, B) is stabilizable and the communication graph $\mathcal{G}$ is connected.*

Proof. If the individual systems reach non-trivial consensus, A is not Hurwitz and the result follows from Corollary 3.8. □

We thus know that graph connectedness, together with stabilizability of the pair (A, B), is necessary for non-trivial consensus in the model (3.4). Note that, for the prototype model (3.1), $A = 0$, i.e., A is not Hurwitz, and non-trivial consensus is reached if and only if the communication graph $\mathcal{G}$ is connected. Though, in contrast to the situation described in Lemma 3.5 for the model (3.1), consensus depends on the feedback gain K for general LTI systems (3.2), and connectedness of the communication graph $\mathcal{G}$ is not sufficient for non-trivial consensus. Therefore, as a next step, we analyze how non-trivial consensus depends on the choice of K and how this choice of K is constrained by the communication topology.

Double Integrator Consensus

We start our analysis with the special case of double integrator systems, i.e., $n = 2, p = 1$, and

$$A = \begin{pmatrix} 0 & 1 \\ 0 & 0 \end{pmatrix}, \quad B = \begin{pmatrix} 0 \\ 1 \end{pmatrix}. \tag{3.9}$$

Consensus among double integrator systems received a lot of attention in the context of multi-agent system consensus. See de Gennaro and Jadbabaie (2006), Olfati-Saber (2006), Ren and Atkins (2005), Ren et al. (2006). The following result is taken from Wieland et al. (2008, 2010a):

Theorem 3.11. *Let $\mathcal{G} = \{\mathcal{V}, \mathcal{E}, W\}$ be a constant communication graph with $|\mathcal{V}| = N > 1$ and Laplacian matrix L satisfying Assumption 2.2. Let $A \in \mathbb{R}^{2\times 2}$ and $B \in \mathbb{R}^{2\times 1}$ be defined as in* (3.9). *Let $K = (k_1, k_2) \in \mathbb{R}^{1\times 2}$.*

The individual systems in the model (3.4) *reach consensus for all initial conditions $x_k(0) \in \mathbb{R}^2$, $k \in \mathbb{N}_N$ if and only if $\mathcal{G}$ is connected and*

$$k_2 < 0, \tag{3.10a}$$

$$\frac{k_2^2}{k_1} < - \max_{j \in \mathbb{N}_N \setminus \{1\}} \frac{1}{\mathfrak{Re}(\lambda_j(L))} \left(\frac{\mathfrak{Im}(\lambda_j(L))}{|\lambda_j(L)|} \right)^2. \tag{3.10b}$$

Proof. Since A is not Hurwitz, by Corollary 3.10 there exist initial conditions such that consensus is not reached if $\mathcal{G}$ is not connected.

Therefore, we need to show that stability of the systems (3.7) is equivalent to conditions (3.10). To establish this equivalence, we apply the Routh-Hurwitz stability criterion (Horn and Johnson, 1991, Theorem 2.3.3) to the matrices

$$\begin{pmatrix} A + \mathfrak{Re}(\lambda_k(L))BK & \mathfrak{Im}(\lambda_k(L))BK \\ -\mathfrak{Im}(\lambda_k(L))BK & A + \mathfrak{Re}(\lambda_k(L))BK \end{pmatrix}$$

for $k \in \mathbb{N}_N \setminus \{1\}$. The resulting stability conditions read

$$\begin{aligned} k_2 &< 0, \\ k_1^2\, \mathfrak{Im}(\lambda_k(L))^2 + k_1 k_2^2\, \mathfrak{Re}(\lambda_k(L))|\lambda_k(L)|^2 &< 0. \end{aligned}$$

The first condition is exactly (3.10a). Note that the second condition implies $k_1 < 0$. Therefore, since $\mathcal{G}$ is connected and thus $\mathfrak{Re}(\lambda_k(L)) > 0$ for all $k \in \mathbb{N}_N \setminus \{1\}$, the second condition is equivalent to

$$\frac{k_2^2}{k_1} < -\frac{1}{\mathfrak{Re}(\lambda_k(L))} \left(\frac{\mathfrak{Im}(\lambda_k(L))}{|\lambda_k(L)|} \right)^2.$$

Since the above condition needs to be satisfied for all $k \in \mathbb{N}_N \setminus \{1\}$, this is equivalent to condition (3.10b). This completes the proof. $\square$

Theorem 3.11 provides necessary and sufficient conditions for consensus among double integrator systems. These conditions relate the elements $k_1, k_2 \in \mathbb{R}$ of the coupling gain matrix K to the spectrum of the Laplacian matrix L of the communication graph $\mathcal{G}$. To obtain the exact bound in (3.10b), complete knowledge of the Laplacian spectrum is required. However, the conditions can be relaxed as follows:

Corollary 3.12. *Let $\mathcal{G} = \{\mathcal{V}, \mathcal{E}, W\}$ be a constant communication graph with $|\mathcal{V}| = N > 1$ and Laplacian matrix L satisfying Assumption 2.2. Let $A \in \mathbb{R}^{2\times 2}$ and $B \in \mathbb{R}^{2\times 1}$ be defined as in* (3.9). *Let $K = (k_1, k_2) \in \mathbb{R}^{1\times 2}$.*

If $\mathcal{G}$ is connected and

$$k_2 < 0, \tag{3.11a}$$

$$\frac{k_2^2}{k_1} < -\frac{1}{\mathfrak{Re}(\lambda_2(L))}, \tag{3.11b}$$

then the individual systems in the model (3.4) *reach consensus for all initial conditions $x_k(0) \in \mathbb{R}^2$, $k \in \mathbb{N}_N$.*

Proof. Since

$$\left(\frac{\mathfrak{Im}(\lambda_k(L))}{|\lambda_k(L)|}\right)^2 \leq 1$$

for any non-zero eigenvalue of L, and $\min_{k\in\mathbb{N}_N\setminus\{1\}} \mathfrak{Re}(\lambda_j(L)) = \mathfrak{Re}(\lambda_2(L))$ by convention, (3.11b) implies (3.10b). This proves the corollary. $\square$

Condition (3.11b) in Corollary 3.12 depends only on the real part of the second eigenvalue of L. It has been argued in Chapter 2 that $\mathfrak{Re}(\lambda_2(L))$ is a measure for the coupling strength of a communication graph $\mathcal{G}$. That is, it suffices to know a lower bound on the coupling strength in order to choose the coupling gain K such that non-trivial consensus is achieved.

One can argue that conditions (3.10) of Theorem 3.11 and conditions (3.11) of Corollary 3.12 represent solutions to the State Consensus Analysis Problem 3.3 as well as the State Consensus Design Problem 3.4 in the case of consensus among double integrator systems. Given a connected communication graph $\mathcal{G}$ and some coupling gain $K = (k_1, k_2) \in \mathbb{R}^{1\times 2}$, those conditions provide an easy method to check whether non-trivial consensus is reached. On the other hand, if the communication graph $\mathcal{G}$ is known to belong to a class of graphs characterized as

$$\mathfrak{G} = \left\{ \mathcal{G} \;\middle|\; \frac{1}{\mathfrak{Re}(\lambda_k(L(\mathcal{G})))}\left(\frac{\mathfrak{Im}(\lambda_k(L(\mathcal{G})))}{|\lambda_k(L(\mathcal{G}))|}\right)^2 \leq a,\ k \in \mathbb{N}_N \setminus \{1\}, a \in \overline{\mathbb{R}}_+ \right\},$$

the conditions $k_2 < 0$ and $-ak_2^2 < k_1 < 0$ provide a simple method to choose the gains k_1 and k_2.

To conclude the discussion of the double integrator case, we would like to give two remarks concerning speed of convergence and a comparison to previously known conditions on the coupling strength for double integrator consensus.

Table 3.1: Different bounds on k_2 with $k_1 = -1$ for non-trivial consensus among double integrators over the communication graph depicted in Figure 3.1.

Theorem 3.11	**Corollary 3.12**	**Ren and Atkins (2005)**
$k_2 < -0.3351$	$k_2 < -0.8646$	$k_2 < -1.223$

Remark 3.13. *Theorem 3.11 and Corollary 3.12 give conditions for non-trivial consensus among double integrator systems. They do not make any statement about the rate of convergence. A slight modification of the argument used in the proof of Theorem 3.11 yields such a statement. If we apply the Routh-Hurwitz stability criterion to the matrices*

$$\begin{pmatrix} A + \mathfrak{Re}(\lambda_k(L))BK & \mathfrak{Im}(\lambda_k(L))BK \\ -\mathfrak{Im}(\lambda_k(L))BK & A + \mathfrak{Re}(\lambda_k(L))BK \end{pmatrix} + \chi I_{2n}$$

for some $\chi \in \mathbb{R}_+$ *and all* $k \in \mathbb{N}_N \setminus \{1\}$, *then* $\sigma(A+\lambda_k(L)BK) \subset \{z \in \mathbb{C} | \mathfrak{Re}(z) \leq -\chi\}$, *i.e.,* χ *determines the exponential rate of convergence. If we set* $k_1 = \chi k_2$, *the corresponding Routh-Hurwitz conditions yield*

$$k_2 < -\frac{2\chi}{\mathfrak{Re}(\lambda_2(L))}.$$

That is, if we choose any $k_2 \in \mathbb{R}$ *satisfying the above inequality and let* $k_1 = \chi k_2$, *then there exists a constant* $M \in \mathbb{R}_+$ *such that* (3.8) *is satisfied with* $\mu = \chi$.

Remark 3.14. *Previous results for double integrator consensus (Ren and Atkins, 2005, Ren et al., 2004) considered a coupling gain* $K = (-1, k_2)$ *and derived the consensus condition*

$$k_2 < -\max_{j\in\mathbb{N}_N\setminus\{1\}} \sqrt{\frac{2}{|\lambda_j(L)| \cos\left(\frac{\pi}{2} - \tan^{-1}\left(\frac{\mathfrak{Re}(\lambda_j(L))}{\mathfrak{Im}(\lambda_j(L))}\right)\right)}}.$$

Some algebraic manipulations lead to the identity

$$\sqrt{\frac{2}{|\lambda_j(L)| \cos\left(\frac{\pi}{2} - \tan^{-1}\left(\frac{\mathfrak{Re}(\lambda_j(L))}{\mathfrak{Im}(\lambda_j(L))}\right)\right)}} = \sqrt{\frac{2}{\mathfrak{Re}(\lambda_j(L))}}.$$

This bound is by a factor $\sqrt{2}$ *worse than the bound given in the sufficient conditions* (3.11) *of Corollary 3.12 in case* $k_1 = -1$. *The exact bounds* (3.10) *given in Theorem 3.11 are significantly tighter. As an example consider the graph depicted in Figure 3.1. The bounds on* k_2 *resulting from Theorem 3.11, Corollary 3.12 and Ren and Atkins (2005) are given in Table 7.1. If the Laplacian spectrum is real, Theorem 3.11 actually predicts that* $k_2 < 0$ *is sufficient for non-trivial consensus independent of the eigenvalues of* L, *while the condition by Ren and Atkins (2005) always yields a bound which depends on the Laplacian spectrum.*

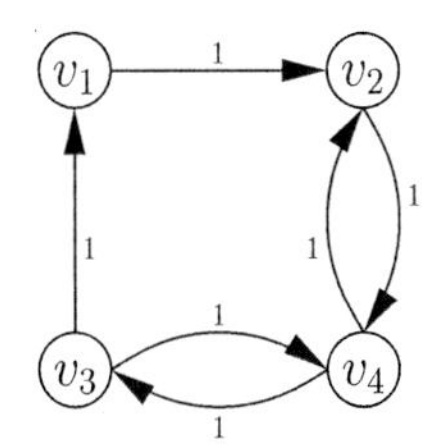

Figure 3.1: Connected graph on 4 vertices where Theorem 3.11 gives significantly better bounds on coupling gains that Ren and Atkins (2005).

A Sufficient Condition for Solvability of the State Consensus Problem

As a consequence of Theorem 3.11, we know that connectedness of the communication graph $\mathcal{G}$ is necessary and sufficient for existence of some coupling gain $K \in \mathbb{R}^{1\times 2}$ such that non-trivial consensus is achieved among double integrator systems. This situation is actually very similar to the situation for the prototype model (3.1) described in Lemma 3.5. In this model, we have $K = -1$. If we considered K as a degree of freedom in the prototype model (3.1), Lemma 3.5 implies that connectedness of the communication graph $\mathcal{G}$ is necessary and sufficient for existence of some coupling gain $K \in \mathbb{R}$ such that non-trivial consensus is achieved in the model (3.1). In fact, non-trivial consensus is achieved for any $K < 0$. The main difference between single and double integrator consensus is that the bounds on the coupling gain K are independent of the Laplacian spectrum for the single integrator case, while they generally depend on the Laplacian spectrum for double integrator consensus as expressed in the inequality (3.10b).

In what follows, we will show that the above statement for consensus among single and double integrator systems generalizes to consensus in the model (3.4) for individuals modeled as arbitrary (but identical) LTI systems. The result below is based on results by Tuna (2008a).

Theorem 3.15. *Let $\mathcal{G} = \{\mathcal{V}, \mathcal{E}, W\}$ be a constant communication graph with $|\mathcal{V}| = N > 1$ and Laplacian matrix L satisfying Assumption 2.2. Let $A \in \mathbb{R}^{n\times n}$ and $B \in \mathbb{R}^{n\times p}$.*

Assume A is not Hurwitz and the pair (A, B) is stabilizable.

Then there exists some matrix $K \in \mathbb{R}^{p\times n}$ such that the individual systems in the model (3.4) reach non-trivial consensus if and only if $\mathcal{G}$ is connected.

Proof. Since, by assumption, A is not Hurwitz we know from Corollary 3.8 that non-trivial consensus cannot be reached if $\mathcal{G}$ is not connected.

It remains to show that connectedness of $\mathcal{G}$ is sufficient for existence of some coupling gain $K \in \mathbb{R}^{p\times n}$ such that the systems (3.7) in Theorem 3.7 are asymptotically stable. We will prove this fact by constructing one such feedback gain K:

Let $\mu \in \mathbb{R}$ with $0 < \mu < 2\,\mathfrak{Re}(\lambda_2(L))$. Such a constant μ exists because connectedness of $\mathcal{G}$ implies $\mathfrak{Re}(\lambda_2(L)) > 0$ by Corollary 2.14. Let $q \in \mathbb{N}$ with $1 \leq q \leq n$ and $M \in \mathbb{R}^{q\times n}$ such that the pair (M, A) is detectable. Define $Q \triangleq M^T M \in \mathbb{R}^{n\times n}$. Let $R \in \mathbb{R}^{p\times p}$ with

$R = R^T \succ 0$. Let $P \in \mathbb{R}^{n \times n}$ be the solution to

$$A^T P + PA - \mu^2 PBR^{-1}B^T P + Q = 0 \tag{3.12}$$

satisfying $P = P^T \succ 0$. Since (A, B) is stabilizable, (M, A) is detectable and $R \succ 0$, this solution is guaranteed to exist and to be unique (see Boyd et al., 1994). We will show that

$$K = -\mu R^{-1} B^T P \tag{3.13}$$

stabilizes all systems (3.7).

Consider the Lyapunov function $V(x) = x^H P x$, $x \in \mathbb{C}^n$. Using (3.13), the time derivative along solutions of (3.7) reads

$$\begin{aligned} \dot{V} &= x^H \left(A^T P + PA - \mu \left(\lambda_k(L) + \overline{\lambda}_k(L)\right) PBR^{-1}B^T P\right) x \\ &= x^H \left(A^T P + PA - 2\mu \,\mathfrak{Re}(\lambda_k(L)) PBR^{-1}B^T P\right) x. \end{aligned}$$

A sufficient condition for stability of systems (3.7) is $\dot{V} < 0$ for $x \neq 0$. With (3.12), this is equivalent to

$$\mu(\mu - 2\,\mathfrak{Re}(\lambda_k(L)))PBR^{-1}B^T P - Q \prec 0.$$

The latter inequality is satisfied for all $k \in \mathbb{N}_N \setminus \{1\}$ because μ was chosen to satisfy $0 < \mu < 2\,\mathfrak{Re}(\lambda_2(L))$. This completes the proof. □

In the above proof, we used the fact that the second eigenvalue $\lambda_2(L)$ of the Laplacian matrix L of a connected graph has positive real part to construct a stabilizable system $\dot{x}(t) = Ax(t) + \mu Bu(t)$. We designed an LQR controller (see Boyd et al., 1994) for that system and used the guaranteed robustness properties of the feedback system $\dot{x}(t) = Ax(t) + \mu Bu(t)$, $y(t) = Kx(t)$ with $u(t) = -y(t)$, which are characterized by a disk margin $D(\frac{1}{2})$ (see Sepulchre et al., 1997). This stability margin guarantees stability of the feedback system with the nominal gain μ replaced by any eigenvalue $\lambda_k(L)$, $k \in \mathbb{N} \setminus \{1\}$ of the Laplacian matrix L, i.e., stability of all systems (3.7) in Theorem 3.7 (see also Wieland et al., 2010a). Consequently, connectedness of $\mathcal{G}$ is sufficient for existence of some coupling gain K that ensures non-trivial consensus. Since connectedness of $\mathcal{G}$ is also necessary as a consequence of Corollary 3.10, we know that the State Consensus Design Problem 3.4 over constant graphs $\mathcal{G}$ has a non-trivial solution if and only if $\mathcal{G}$ is connected. Prior to continuing with the discussion of the State Consensus Design Problem 3.4, a short remark is in order concerning an important generalization of the above ideas:

Remark 3.16. *Exploiting the close link between disk margins and passivity (Sepulchre et al., 1997), similar ideas can be used to show consensus and synchronization among passive systems as, e.g., in Arcak (2007), Chopra and Spong (2006), Pogromsky (1997), Stan and Sepulchre (2007); we do not pursue these ideas any further in this thesis.*

In contrast to consensus among double integrator systems, we do not have exact bounds on the coupling gain K for general LTI systems. However, the proof of Theorem 3.15 provides a design method for K that only depends on a lower bound on $\mathfrak{Re}(\lambda_2(L))$, similarly to Corollary 3.12 in the case of double integrator systems. Subsequently, an alternative solution to the State Consensus Design Problem 3.4 is presented.

An LMI-based Solution to the State Consensus Design Problem

The results below have been published in Wieland et al. (2008, 2010a).

Theorem 3.17. *Let $\mathcal{G} = \{\mathcal{V}, \mathcal{E}, W\}$ be a constant communication graph with $|\mathcal{V}| = N > 1$ and Laplacian matrix L satisfying Assumption 2.2. Let $A \in \mathbb{R}^{n\times n}$, $B \in \mathbb{R}^{n\times p}$, and $\chi \in \mathbb{R}_+$.*
If there exist matrices $Q \in \mathbb{R}^{n\times n}$ and $\kappa \in \mathbb{R}^{p\times n}$ such that $Q = Q^T \succ 0$ and

$$QA^T + AQ + 2\chi Q + \lambda_k(L)B\kappa + \overline{\lambda}_k(L)\kappa^T B^T \preceq 0, \quad k \in \mathbb{N}_N \setminus \{1\}, \tag{3.14}$$

then the coupling gain $K = \kappa Q^{-1}$ in (3.4b) ensures that there exists a constant $M \in \mathbb{R}_+$ such that (3.8) is satisfied with $\mu = \chi$.

Proof. Let $P = Q^{-1}$. Then substituting (3.13) into (3.12) and multiplying by P from the left and the right yields

$$(A + \lambda_k(L)BK)^H P + P(A + \lambda_k(L)BK) \preceq -2\chi P,\ k \in \mathbb{N}_N \setminus \{1\}$$

Together with $P = P^T \succ 0$, the above matrix inequality proves exponential stability of the systems (3.7) with exponential convergence rate χ (see Boyd and Vandenberghe, 2004, Boyd et al., 1994, for details). In fact, $V(x) = x^H Px$, $x \in \mathbb{C}^n$ is a quadratic Lyapunov function for the systems (3.7) with $\dot{V} \leq -2\chi V$. Therefore

$$\|x^H(t)Px(t)\| \leq e^{-2\chi t}\|x^H(0)Px(0)\|$$

and with $\lambda_1(P)\|x\|^2 \leq \|x^H Px\| \leq \lambda_n(P)\|x\|^2$ for any $x \in \mathbb{C}$, this yields

$$\|x(t)\| \leq \sqrt{\frac{\lambda_n(P)}{\lambda_1(P)}}e^{-\chi t}\|x(0)\|$$

for any solution $x(t)$ of the systems (3.7). □

While existence of a quadratic Lyapunov function is necessary and sufficient for asymptotic stability of an LTI system, the conditions in Theorem 3.17 are conservative to some extent. Firstly, since the systems (3.7) may be complex due to the complex gains $\lambda_k(L)$, restricting Q to be real symmetric instead of complex Hermitian may be conservative. Secondly, and more importantly, we use the same Lyapunov function $x^H Px$ to prove asymptotic stability of all systems (3.7) for $k \in \mathbb{N}_N \setminus \{1\}$. This second constraint is necessary in order to obtain a *linear* matrix inequality. Allowing for different Lyapunov functions for the different systems would result in a *bilinear* matrix inequality involving terms $P_k BK$, $k \in \mathbb{N}_N \setminus \{1\}$ with unknowns P_k, $k \in \mathbb{N}_N \setminus \{1\}$ and K. However, if P is the unique positive definite solution to (3.12) and K is defined as (3.13), we know that $Q = P^{-1}$ and $\kappa = KQ$ solve (3.14). That is, if $\mathcal{G}$ is connected and χ is chosen small enough (or more precisely if χ is such that the pair $(A + \chi I_n, B)$ is stabilizable), there always exists a solution to the LMI (3.14) despite the conditions being conservative.

A drawback of the LMI condition in Theorem 3.17 above is the explicit dependence on the complete spectrum of the Laplacian matrix L (except for the 0 eigenvalue). Exploiting convexity of (3.14) in $\lambda_k(L)$, we give relaxed conditions below, that are easier to check based on simple bounds on the Laplacian spectrum.

Corollary 3.18. *Let $\mathcal{G} = \{\mathcal{V}, \mathcal{E}, W\}$ be a constant communication graph with $|\mathcal{V}| = N > 1$ and Laplacian matrix L satisfying Assumption 2.2. Let $A \in \mathbb{R}^{n\times n}$, $B \in \mathbb{R}^{n\times p}$, and $\chi \in \mathbb{R}_+$. Let $r \in \mathbb{N}$ and $\mu_j \in \mathbb{C}$, $j \in \mathbb{N}_r$ such that $\lambda_k(L) \in \operatorname{conv}(\{\mu_1, \dots, \mu_r, \overline{\mu}_1, \dots, \overline{\mu}_r\}) \subset \mathbb{C}$ for all $k \in \mathbb{N}_N \setminus \{1\}$.*

If there exist matrices $Q \in \mathbb{R}^{n\times n}$ and $\kappa \in \mathbb{R}^{p\times n}$ such that $Q = Q^T \succ 0$ and

$$QA^T + AQ + 2\chi Q + \mu_j B\kappa + \overline{\mu}_j \kappa^T B^T \preceq 0, \quad j \in \mathbb{N}_r, \tag{3.15}$$

then the coupling gain $K = \kappa Q^{-1}$ in (3.4b) ensures that there exists a constant $M \in \mathbb{R}_+$ such that (3.8) is satisfied with $\mu = \chi$.

Proof. We define the matrix-valued map $C : \mathbb{C} \to \mathbb{C}^{n\times n}$ as $C(\mu) = QA^T + AQ + 2\chi Q + \mu B\kappa + \overline{\mu}\kappa^T B^T$. Since $C^T(\mu) = C(\overline{\mu})$, $C(\mu) \preceq 0$ is equivalent to $C(\overline{\mu}) \preceq 0$. Furthermore $C(\mu_1) \preceq 0$ and $C(\mu_2) \preceq 0$ implies that $C(\theta\mu_1 + (1-\theta)\mu_2) \preceq 0$ for all $\theta \in [0,1]$. Thus the set $\{\mu \in \mathbb{C} | C(\mu) \preceq 0\}$ is convex and symmetric with respect to the real axis.

Consequently, (3.15) implies that $C(\mu) \preceq 0$ for any $\mu \in \operatorname{conv}(\{\mu_1, \dots, \mu_r, \overline{\mu}_1, \dots, \overline{\mu}_r\})$. Therefore, by choice of the complex numbers μ_j, $j \in \mathbb{N}_r$, (3.14) follows. □

With the help of Corollary 3.18, we are now able to design the coupling gain K with the help of the knowledge of any convex set containing the non-zero eigenvalues of the Laplacian matrix L. However, this is of limited use, as long as we are not able to construct such a convex set based on limited knowledge on the Laplacian spectrum. The corollary below is based on one possibility to construct a convex set containing all non-zero eigenvalues of L.

Corollary 3.19. *Let $\mathcal{G} = \{\mathcal{V}, \mathcal{E}, W\}$ be a constant communication graph with $|\mathcal{V}| = N > 1$ and Laplacian matrix L satisfying Assumption 2.2. Let $A \in \mathbb{R}^{n\times n}$, $B \in \mathbb{R}^{n\times p}$, and $\chi \in \mathbb{R}_+$. Let $\sigma, d \in \mathbb{R}_+$ with $\sigma \le \mathfrak{Re}(\lambda_2(L))$ and $d \ge \max_k \sum_{j=1}^N w_{k,j}$. Let $\mu_1 = \sigma + \mathrm{j}\sigma\sqrt{(2d-\sigma)/\sigma}$, $\mu_2 = d\sqrt{\sigma/(2d-\sigma)} + \mathrm{j}d$, and $\mu_3 = 2d + \mathrm{j}d$.*

If there exist matrices $Q \in \mathbb{R}^{n\times n}$ and $\kappa \in \mathbb{R}^{p\times n}$ such that $Q = Q^T \succ 0$ and (3.15) is satisfied for all $j \in \mathbb{N}_3$, then the coupling gain $K = \kappa Q^{-1}$ in (3.4b) ensures that there exists a constant $M \in \mathbb{R}_+$ such that (3.8) is satisfied with $\mu = \chi$.

Proof. We shall show that $\lambda_k(L) \in \operatorname{conv}(\{\mu_1, \mu_2, \mu_3, \overline{\mu}_1, \overline{\mu}_2, \overline{\mu}_3\})$, $k \in \mathbb{N}_N \setminus \{1\}$. To this end, we define a set $\mathcal{L}(d, \sigma) \subset \mathbb{C}$ as

$$\mathcal{L}(d,\sigma) \triangleq \{\lambda \in \mathbb{C} : |\lambda - d| \le d \wedge \mathfrak{Re}(\lambda) \ge \sigma\}.$$

Let $d_{\min} = \max_{k\in\mathbb{N}_N} \sum_{j=1}^N w_{k,j}$ and $\sigma_{\max} = \mathfrak{Re}(\lambda_2(L))$. As a consequence of the the Geršgorin Disk Theorem (see Horn and Johnson, 1985, Theorem 6.1.1), we know that the eigenvalues of the Laplacian matrix satisfy $\lambda_k(L) \in \mathcal{L}(d_{\min}, \sigma_{\max})$, $k \in \mathbb{N}_N \setminus \{1\}$. Furthermore, by choice of d and σ, we have that $\mathcal{L}(d_{\min}, \sigma_{\max}) \subseteq \mathcal{L}(d, \sigma)$. Since the complex number μ_j, $j \in \mathbb{N}_3$ have been chosen such that $\mathcal{L}(d,\sigma) \subset \operatorname{conv}(\{\mu_1, \mu_2, \mu_3, \overline{\mu}_1, \overline{\mu}_2, \overline{\mu}_3\})$, $\lambda_k(L) \in \operatorname{conv}(\{\mu_1, \mu_2, \mu_3, \overline{\mu}_1, \overline{\mu}_2, \overline{\mu}_3\})$, $k \in \mathbb{N}_N \setminus \{1\}$ follows. The different sets used in the proof are illustrated in Figure 3.2. □

The numbers $\mu_j \in \mathbb{C}$, $j \in \mathbb{N}_3$ given in Corollary 3.19 depend on a lower bound for the real part of the second eigenvalue $\lambda_2(L)$ of the Laplacian matrix L, similarly to Corollary 3.12

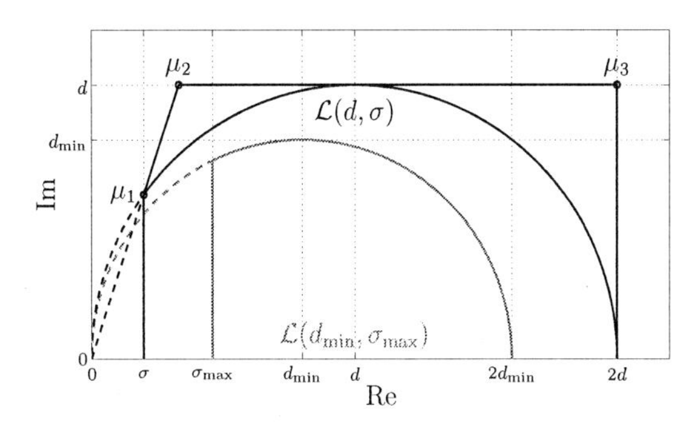

Figure 3.2: Construction of convex sets containing the non-zero eigenvalues of the Laplacian matrix L

for consensus among double integrators. In addition, the numbers μ_j depend and on an upper bound on the maximal diagonal element of L or equivalently an upper bound on the maximal edge weight with the relation $\max_{k \in \mathbb{N}_N} \sum_{j=1}^{N} w_{k,j} \leq (N-1) \max_{j,k \in \mathbb{N}_N} w_{k,j}$. Such a number is frequently known since edge weights are commonly bounded by design, e.g., when edge weights are constraint to take values in $\{0, 1\}$.

Example

We will illustrate the results obtained for consensus among general LTI systems over fixed communication graphs on a formation problem for vehicles moving on a plane. We assume a group of holonomic vehicles with identical dynamics whose objective is to reach a fixed formation. Each vehicle in the group is assigned a fixed and a priori known position within the target formation. Thus, the formation problem is readily reduced to a consensus problem.

In the literature, very often kinematic models, i.e., integrator models, or double integrator models are employed in vehicle formation problems (see Olfati-Saber, 2006, Ren et al., 2007, and the references therein). This approach however neglects damping and actuator dynamics. In the present example, we consider holonomic vehicles whose models include both damping and actuator dynamics. The N vehicles are modeled as

$$\begin{aligned} \dot{y}_k(t) &= v_k(t) \\ \dot{v}_k(t) &= -\alpha_1 v_k(t) + p_k(t) \qquad k \in \mathbb{N}_N, \\ \dot{p}_k(t) &= -\alpha_2 p_k(t) + w_k(t) \end{aligned}$$

where $y_k \in \mathbb{R}^2$ is the position on the plane, $v_k \in \mathbb{R}^2$ represents the kth vehicle's velocity, $p_k \in \mathbb{R}^2$ is an actuator state, and $w_k \in \mathbb{R}^2$ is an input. The parameters $\alpha_1, \alpha_2 \in \mathbb{R}_+$ characterize the damping and the actuator dynamics. The target position of the kth vehicle relative to the center of the formation is given by some constant vector $d_k \in \mathbb{R}^2$.

Because the vehicles are considered to be holonomic, the vehicles evolve independently on the two planar directions. For that reason, we only solve the consensus problem in one

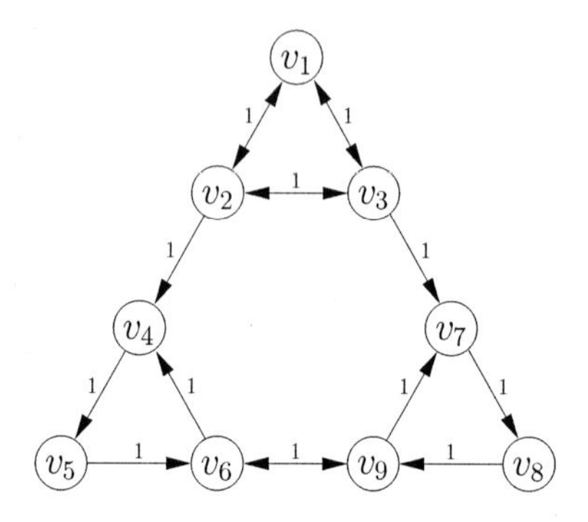

Figure 3.3: Shape of target formation and communication graph.

direction resulting in the individual system model

$$\dot{x}_k(t) = \begin{pmatrix} 0 & 1 & 0 \\ 0 & -\alpha_1 & 1 \\ 0 & 0 & -\alpha_2 \end{pmatrix} x_k(t) + \begin{pmatrix} 0 \\ 0 \\ 1 \end{pmatrix} u_k(t), \qquad k \in \mathbb{N}_N \tag{3.16}$$

with state vector $x_k(t) = (y_{k,j}(t) - d_{k,j}, v_{k,j}(t), p_{k,j}(t))^T \in \mathbb{R}^3$ and input $u_k(t) = w_{k,j}(t) \in \mathbb{R}$ for $j = 1$ or $j = 2$. The coordinate $y_{k,j}(t) - d_{k,j}$ corresponds to the position of the kth system translated to the center of the target formation. That is, if a consensus for the state vectors $x_k(t)$, $k \in \mathbb{N}_N$ is achieved, the vehicles have reached the desired formation. If $\xi(t)$ is a consensus trajectory for the systems (3.16), then $y_{k,j}(t) \to \zeta_{k,j}(t)$, $k \in \mathbb{N}_9$ for $j = 1$ or $j = 2$ as $t \to \infty$, where $\zeta_{k,j}(t) \triangleq \xi_1(t) + d_{k,j}$ is an *individual consensus trajectory* for the kth system.

The target formation and the communication graph $\mathcal{G} = \{\mathcal{V}, \mathcal{E}, W\}$ are depicted in Figure 3.3. The graph $\mathcal{G}$ contains one iSCC consisting of the vertices v_1, v_2, and v_3 and is thus connected. The second eigenvalue of the corresponding Laplacian matrix L is $\lambda_2(L) \approx 0.2451$.

With the parameters in the vehicle dynamics chosen as $\alpha_1 = 1$ and $\alpha_2 = 5$, Corollary 3.19 with $\sigma = 0.2$, $d = (N-1) = 8$, and $\chi = 2$ yields $K \approx -(284.1, 154.4, 25.38)$. For reference, we also give the solution to the LMIs (3.14) in Theorem 3.17 with $\chi = 2$ in which case the solution $K \approx -(310.8, 157.8, 21.42)$ has been found. The fact that the coupling gains resulting from Corollary 3.19 are smaller than the coupling gains obtained from Theorem 3.17 is not a generic property of the proposed methods but is rather due to implementation details of the LMI solver that has been used. Simulation results are depicted in Figure 3.4. Since the system dynamics (3.16) are stable, the individuals eventually reach constant positions corresponding to the target formation. This consensus position depends only on the initial conditions of the systems represented by vertices v_1, v_2, and v_3, i.e., the members of the unique iSCC of the communication graph $\mathcal{G}$.

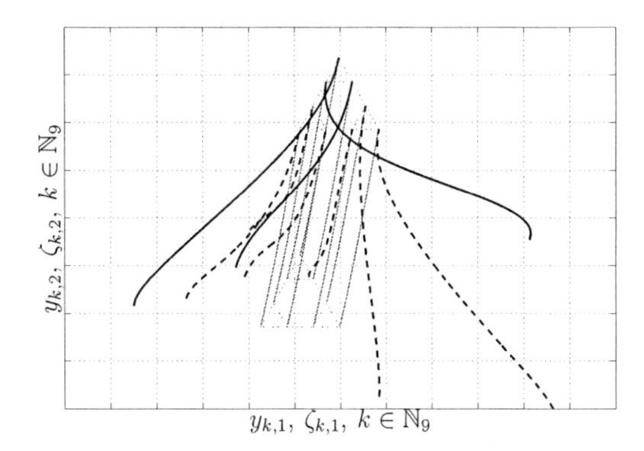

Figure 3.4: Vehicle positions $y_{k,j}$, $k \in \mathbb{N}_9$, $j \in \{1, 2\}$ in the plane (black lines) for iSCC members (solid) and non iSCC members (dashed) and individual consensus trajectories $\zeta_{k,j}$, $k \in \mathbb{N}_9$, $j \in \{1, 2\}$ (grey lines).

3.3 Consensus for Switching and Time-Varying Communication Topologies

In the previous section, we were concerned with consensus of arbitrary LTI systems over fixed communication graphs. In many cases, the assumption of a constant communication topology is however invalid. In the Vicsek model (Vicsek et al., 1995), the communication topology is determined by nearest neighbor rules, i.e., what an individual 'sees' depends on where it is located in space. In gossip algorithms (Boyd et al., 2005), individual systems are often considered close to each other, if the values to agree upon are already close to each other. This reflects the fact that individuals tend to have higher confidence in beliefs that are close to their own belief. As a result, the communication topology may depend directly on the value consensus shall be achieved about.

Very often, the exact evolution of the communication graph as a function of time is unknown or hard to predict. The results below provide thus necessary conditions for consensus over time-varying communication graphs that do not depend on a specific evolution of the communication graph but rather on uniform connectedness in the sense of Definition 2.6. These results apply to situations where the communication graph depends on the states to agree upon as well as situations where the communication graph depends on exogenous signals.

The first version of the result applies to the prototype model (3.1). It is taken from Moreau (2004a,b, 2005) and will be stated below without proof. The proof can be found in the original publications cited before.

Theorem 3.20. *Let $\mathcal{G}(t) = \{\mathcal{V}, \mathcal{E}(t), W(t)\}$ be a communication graph with $|\mathcal{V}| = N > 1$ and Laplacian matrix $L(t)$.*

If $\mathcal{G}(t)$ is uniformly connected, then consensus in the prototype model (3.1) *is reached for all initial conditions $x(0) \in \mathbb{R}^N$ and there exist constants $M, \mu \in \mathbb{R}_+$ and a vector*

Table 3.2: Comparison between Lemma 3.5 and Theorem 3.20.

	Lemma 3.5 (constant graphs)	**Theorem 3.20** (time-varying graphs)
Type of Condition	connectedness is *necessary and sufficient*	uniform connectedness is *only sufficient*
Consensus Value	Consensus value is given as $p^T x(0)$, where $p^T L = 0$ and $p^T 1_N = 1$	Consensus value is contained in $[\min_{k\in\mathbb{N}_N} x_k(0), \max_{k\in\mathbb{N}_N} x_k(0)]$ but it is not possible to predict the exact consensus value
Convergence Rate	Bounded by $\mathfrak{Re}(\lambda_2(L))$	Bounded as a function of the threshold α in Definition 2.1 and the time horizon T in Definition 2.6; may be arbitrarily slow

$\xi_0 \in [\min_{k\in\mathbb{N}_N} x_k(0), \max_{k\in\mathbb{N}_N} x_k(0)] \subset \mathbb{R}$ *such that*

$$\|x(t) - 1_N\xi_0\| \leq Me^{-\mu t}\|x(0) - 1_N\xi_0\|, \tag{3.17}$$

i.e., the consensus value, given by ξ_0, *is reached exponentially fast.*

The idea of the proof of Theorem 3.20 in Moreau (2004a,b, 2005) is based on the observation that on the one hand the network cannot diverge when not connected and on the other hand connected parts of the network exponentially converge. Uniform connectedness of the communication graph ensures that any two distinct vertices in the graph can share information from time to time. Using these observations, exponential convergence to consensus is established. Similar ideas have been proposed earlier by Tsitsiklis (1984).

In contrast to Lemma 3.5, where the communication graph was assumed to be constant, Theorem 3.20 allows for arbitrary time-varying communication graphs $\mathcal{G}$ as long as they are uniformly connected. The time horizon T in Definition 2.6 may be arbitrarily large. However, speed of convergence generally decreases for increasing T. Besides strongly relaxed connectedness conditions, i.e., the possibility to deal with significantly higher topological complexity, there are some more remarkable properties of Theorem 3.20 as compared to Lemma 3.5. A comparison of the two results is given in Table 3.2.

As in the case of constant communication graphs, there exists a simple corollary to Theorem 3.20 providing a condition for average consensus.

Corollary 3.21. *Let* $\mathcal{G}(t) = \{\mathcal{V}, \mathcal{E}(t), W(t)\}$ *be a communication graph with* $|\mathcal{V}| = N > 1$ *and Laplacian matrix* $L(t)$.

If $\mathcal{G}(t)$ *is uniformly connected and balanced at all times, then average consensus in the prototype model* (3.1) *is reached for all initial conditions* $x(0) \in \mathbb{R}^N$ *and there exist constants* $M, \mu \in \mathbb{R}_+$ *such that*

$$\left\|x(t) - 1_N\frac{1_N^T x(0)}{N}\right\| \leq Me^{-\mu t}\left\|x(0) - 1_N\frac{1_N^T x(0)}{N}\right\|, \tag{3.18}$$

i.e., average-consensus is reached exponentially fast.

Proof. Consensus is exponentially reached by Theorem 3.20. We need to show that the consensus value is the average of the initial conditions. Let $\xi(t) \triangleq (1_N^T x(t))/N \in \mathbb{R}$ be the average of the states $x_k(t)$, $k \in \mathbb{N}_N$ at time t. Then $\dot{\xi}(t) = 1_N L(t)x(t)$. Since $\mathcal{G}(t)$ is balanced at all times, $1_N L(t) = 0$ at all times. Thus $\xi(t) \equiv (1_N^T x(0))/N$. Thus, if consensus is reached, then $x_k(t) \to (1_N^T x(0))/N$ as $t \to \infty$. □

In Section 3.2, we generalized the result of Lemma 3.5 for consensus in the prototype model (3.1) to general LTI systems (3.4). We provided solutions to the State Consensus Analysis Problem 3.3 as well as the State Consensus Design Problem 3.4. The question to be answered in what follows is to what extent similar results exist for time-varying communication topologies. One answer to that question has been given in Scardovi and Sepulchre (2008, 2009) and is repeated in the following theorem:

Theorem 3.22. *Let $\mathcal{G}(t) = \{\mathcal{V}, \mathcal{E}(t), W(t)\}$ be a communication graph with $|\mathcal{V}| = N > 1$ and Laplacian matrix $L(t)$. Let $A \in \mathbb{R}^{n\times n}$, $B = I_n$, and $K = -I_n$.*

If $\mathcal{G}(t)$ is uniformly connected and $\sigma(A) \subset \overline{\mathbb{C}}_-$, then consensus in the model (3.4) is reached for all initial conditions $x(0) \in \mathbb{R}^{Nn}$ and there exist constants $M, \mu \in \mathbb{R}_+$ and a vector $\xi_0 \in \operatorname{conv}(\{x_1(0), \ldots, x_N(0)\}) \subset \mathbb{R}^n$ such that

$$\left\| x(t) - 1_N \otimes \left(e^{At}\xi_0\right)\right\| \leq Me^{-\mu t} \left\| x(0) - 1_N \otimes \xi_0 \right\|, \quad t \geq 0, \tag{3.19}$$

i.e., the consensus trajectory, given by $\xi(t) = e^{At}\xi_0$, is reached exponentially fast.

Proof. Consider the change of variables $z_k(t) = e^{-At}x_k(t)$, $k \in \mathbb{N}_N$. Then

$$\begin{aligned}\dot{z}_k(t) &= -Ae^{-At}x(t) + e^{-At}Ax(t) - e^{-At}\sum_{j=1}^{N} w_{k,j}(t)(x_k(t) - x_j(t)) \\ &= -\sum_{j=1}^{N} w_{k,j}(t)(z_k(t) - z_j(t)).\end{aligned}$$

That is, the quantities $z_{q,k}(t) \triangleq q^T z_k(t)$, $k \in \mathbb{N}_N$ satisfy $\dot{z}_{q,k}(t) = -\sum_{j=1}^{N} w_{k,j}(t)(z_{q,k}(t) - z_{q,j}(t))$ and thus reach consensus with consensus value in the interval $[\min_{k\in\mathbb{N}_N} z_{q,k}(0), \max_{k\in\mathbb{N}_N} z_{q,k}(0)]$ by Theorem 3.20 for any vector $q \in \mathbb{R}^n$. Thus, there exist constants $M_1, \mu_1 \in \mathbb{R}_+$ and $\xi_0 \in \operatorname{conv}(\{z_1(0), \ldots, z_N(0)\}) \subset \mathbb{R}^n$ such that

$$\|z(t) - 1_N \otimes \xi_0\| \leq M_1 e^{-\mu_1 t}\|z(0) - 1_N \otimes \xi_0\|.$$

Since $\sigma(A) \subset \overline{\mathbb{C}}_-$, we know that $\left\|e^{At}\right\| \leq p(t)$ for some real polynomial $p(t)$ of degree less than or equal to n in t satisfying $p(t) \geq 0$ for $t \geq 0$. Using $x(0) = z(0)$, it follows that for $t \geq 0$,

$$\begin{aligned}\left\|x(t) - 1_N \otimes \left(e^{At}\xi_0\right)\right\| &= \left\|\left(I_N \otimes e^{At}\right)\left(z(t) - 1_N \otimes \xi_0\right)\right\| \\ &\leq \left\|I_N \otimes e^{At}\right\| \left\|\left(z(t) - 1_N \otimes \xi_0\right)\right\| \\ &\leq p(t)M_1 e^{-\mu_1 t} \left\|x(0) - 1_N \otimes \xi_0\right\|.\end{aligned}$$

Thus, for any $\mu < \mu_1$, there exists a constant $M \in \mathbb{R}_+$ such that (3.19) holds. □

Table 3.3: Comparison between Theorem 3.17 and Theorem 3.22.

	Theorem 3.17 (constant graphs)	**Theorem 3.22** (time-varying graphs)
Communication Topology	$\mathcal{G}$ connected	$\mathcal{G}(t)$ uniformly connected
Coupling gain	K is degree of freedom; knowledge of Laplacian spectrum required	$K = -I$ (or $K = -B^{-1}$)
Dynamics	arbitrary	$\sigma(A) \subset \overline{\mathbb{C}}_-$
Inputs	(A, B) stabilizable	$B = I$ (or $\text{rank}(B) = n$)

It is interesting to compare Theorem 3.22 with the results for fixed communication topologies presented in Section 3.2, in particular Theorem 3.17. Some of the features of these two results are compared in Table 3.3. As argued before, Theorem 3.22 allows for significantly higher topological complexity. But there is a price to pay with respect to system complexity. Theorem 3.17 applies to individual systems modeled with arbitrary dynamics matrices A and more importantly only requires (A, B) to be stabilizable, while the result in this section requires $p = n$ and $B = I_n$[2], i.e., thus full control of all the states of the individual systems.

Particularly the requirement of having full control of all states may be restrictive in practice. Theorem 3.22 puts this requirement as an assumption, i.e., $B = I_n$ is not a necessary condition by Theorem 3.22. But simple examples have been reported in Scardovi and Sepulchre (2008, 2009), where $\text{rank}(B) < n$ prevents individual systems from reaching a consensus over uniformly connected communication graphs. This suggests that there is some trade-off between topological complexity and system complexity for consensus with static diffusive couplings.

3.4 Consensus with Leader

In this section, we will have a look at the special case of consensus in presence of a leader or a reference. A leader is one particular individual system which prescribes the asymptotic behavior of the whole network of systems, i.e., all individuals adopt the behavior of the leader or reference. Examples of such problems can be found for instance in Bai et al. (2009), Ren (2007), Tanner et al. (2004).

In this section, we will first consider the case that the leader is modeled with dynamics identical to the dynamics of the other network members. As a second step, we consider the case, when the leader dynamics differ from the dynamics of the other network members. Finally, we will comment on the problem of connecting a leader to an existing network of dynamical systems, i.e., the identification of those systems in a group which a leader needs to influence such that the whole group asymptotically follows the leader.

[2]The requirement $B = I_n$ could be replaced by $\text{rank}(B) = n$. In that case $K = -I_n$ can be replaced by $K = -B^{-1}$ to obtain the same result.

3.4.1 Leader with Identical Dynamics

As a consequence of the symmetry inherent to the diffusive couplings (3.4b), consensus with a leader is readily dealt with in the framework of diffusively coupled systems. In fact, consensus with a leader whose dynamics correspond to the dynamics of the other systems in the network merely corresponds to a specific communication topology, as stated below.

Lemma 3.23. *Let $\mathcal{G}(t) = \{\mathcal{V}, \mathcal{E}(t), W(t)\}$ be a communication graph with $|\mathcal{V}| = N > 1$ and Laplacian matrix $L(t)$. Let $A \in \mathbb{R}^{n\times n}$, $B \in \mathbb{R}^{n\times p}$, and $K \in \mathbb{R}^{p\times n}$.*

Assume the individual systems in the model (3.4) reach consensus independent of initial conditions $x_k(0) \in \mathbb{R}^n$, $k \in \mathbb{N}_N$.

If there exists some $k \in \mathbb{N}_N$ such that $w_{k,j}(t) \equiv 0$, $j \in \mathbb{N}_N$, then $\xi(t) = e^{At}x_k(0)$ is a consensus trajectory, i.e., the kth system is a leader.

Proof. If $w_{k,j}(t) \equiv 0$, $j \in \mathbb{N}_N$, then $u_k(t) \equiv 0$, i.e., $x_k(t) = e^{At}x_k(0)$. Then consensus implies

$$\lim_{t\to\infty} \left\|x_j(t) - e^{At}x_k(0)\right\| = 0,\ j \in \mathbb{N}_N.$$

This completes the proof. □

The condition that $w_{k,j}(t) \equiv 0$, $j \in \mathbb{N}_N$ implies that the system represented by vertex $v_k \in \mathcal{V}$ evolves independently of all other systems uniformly in t. If the graph $\mathcal{G}(t)$ satisfies Assumption 2.2, this is equivalent to the condition that there is no edge in $\mathcal{E}(t)$ with v_k as its head uniformly in t, i.e., $\{v_j \in \mathcal{V} : (v_j, v_k) \in \mathcal{E}(t)\} = \emptyset$ for all $t \in \mathbb{R}$[3]. Another way to state this property is to say that the graph $\tilde{\mathcal{G}}$ induced by $\tilde{\mathcal{V}} \triangleq \{v_k\} \subset \mathcal{V}$ must be an iSCC of $\mathcal{G}(t)$ uniformly in t. If $\mathcal{G}(t) = \mathcal{G}$ is constant and satisfies Assumption 2.2, the condition in Lemma 3.23 is equivalent to the condition that $\tilde{\mathcal{G}}$ is the unique iSCC of $\mathcal{G}$ as a consequence of Theorem 2.13.

3.4.2 Leader with Different Dynamics

Lemma 3.23 above applies for constant and time-varying communication topologies. The leader is simply one of the individual members of the network. It is qualified as a leader or reference merely through specific properties of the communication topology. As a consequence, the leader essentially determines the initial condition $x_k(0) \in \mathbb{R}^n$ of the solution $x_k(t)$ to the open loop system (3.2) with $u_k(t) \equiv 0$ which serves as a consensus trajectory. One may ask the question, whether it is also possible to influence the dynamic evolution of the group at consensus through a leader. This question is addressed in the following.

In what follows, we consider a leader in addition to a network of N individual systems. The leader is modeled as an autonomous LTI system

$$\dot{\nu}(t) = S\nu(t) \tag{3.20}$$

with state vector $\nu(t) \in \mathbb{R}^n$ of the same dimension as the state vectors $x_k(t)$ of the individual systems (3.2). Given some initial condition $\nu(0) \in \mathbb{R}^n$, consensus with the

[3]The implication $w_{k,j}(t) = 0 \Rightarrow (v_j, v_k) \notin \mathcal{E}(t)$ is always true. Assumption 2.2 ensures that $w_{k,j}(t) = 0 \Leftarrow (v_j, v_k) \notin \mathcal{E}(t)$.

leader (3.20) means that $\xi(t) = \nu(t) = e^{St}\nu(0)$ is a consensus trajectory. The couplings (3.3) are modified to account for the leader as

$$u_k(t) = K\sum_{j=1}^{N} w_{k,j}(t)(x_k(t) - x_j(t)) + E_1 c_k(t)(x_k(t) - \nu(t)) + E_2 c_k(t)\nu(t) \qquad (3.21)$$

for all $k \in \mathbb{N}_N$. The matrices $E_1, E_2 \in \mathbb{R}^{p\times n}$ are additional design parameters and $c_k(\cdot) : t \to \overline{\mathbb{R}}_+$, $k \in \mathbb{N}$ are piecewise continuous functions of time whose value at some specific time indicates whether there exists a link between the leader and the kth system or not at that time. The above couplings include the leader in a very general way, allowing for a relative feedback part with gain E_1 and a feedforward part with gain E_2. We will later on comment on the role of these two parts and put constraints on the values that E_1 and E_2 may take.

Let $C(t) = \operatorname{diag}\left((c_1(t), \ldots, c_N(t))^T\right) \in \mathbb{R}^{N\times N}$ and define the error variable $e(t) = x(t) - (1_N \otimes \nu(t)) \in \mathbb{R}^{Nn}$. Then (3.2) and (3.21) yield

$$\dot{e}(t) = (I \otimes A)e(t) + (I \otimes (A - S))(1_N \otimes \nu(t)) + (I \otimes B)u(t) \qquad (3.22a)$$

$$u(t) = \left[(L(t) \otimes K) + (C(t) \otimes E_1)\right] e(t) + (C(t) \otimes E_2)(1_N \otimes \nu(t)), \qquad (3.22b)$$

where we used the fact that $(L(t) \otimes K)e(t) = (L(t) \otimes K)x(t)$. Since $e(t)$ is the consensus error, $\nu(t)$ is a consensus trajectory if and only if $e(t) \to 0$ as $t \to \infty$. We use this fact to establish a necessary condition for consensus of the model (3.22) with leader (3.20) below (see Wieland et al., 2010a).

Theorem 3.24. *Let $\mathcal{G}(t) = \{\mathcal{V}, \mathcal{E}(t), W(t)\}$ be a communication graph with $|\mathcal{V}| = N > 1$ and Laplacian matrix $L(t)$. Let $A \in \mathbb{R}^{n\times n}$, $B \in \mathbb{R}^{n\times p}$, $K \in \mathbb{R}^{p\times n}$, $S \in \mathbb{R}^{n\times n}$, and $E_1, E_2 \in \mathbb{R}^{p\times n}$. Let $c_k(\cdot) : \mathbb{R} \to \overline{\mathbb{R}}_+$, $k \in \mathbb{N}_N$ be piecewise continuous functions of time. Let $\mathcal{T} \subset \mathbb{R}^n$ be the invariant subspace for the leader dynamics (3.20) which is maximal subject to not containing asymptotically stable modes and let $T \in \mathbb{R}^{n\times q}$ be such that $\operatorname{rank}(T) = q \in \mathbb{N}_{ns} \bigcup\{0\}$ and $\operatorname{im}(T) = \mathcal{T}$.*

If the individual systems in the model (3.22) reach consensus for all initial conditions $\nu(0) \in \mathbb{R}^n$, then

$$\lim_{t\to\infty} c_k(t)BE_2T = (S - A)T, \quad k = 1, \ldots, N. \qquad (3.23)$$

Proof. We already explained before that consensus in the model (3.22) with leader (3.20) is equivalent to $e(t) \to 0$ as $t \to \infty$. The model (3.22) represents a non-autonomous dynamical system with state $e(t)$ and input $\nu(t)$. A necessary condition for $e(t) \to 0$ as $t \to \infty$ is thus that the influence of $\nu(t)$ in (3.22) asymptotically vanishes, i.e.,

$$\lim_{t\to\infty} \left[(I \otimes (A - S)) + (C(t) \otimes BE_2)\right](1_N \otimes \nu(t)) = 0$$

We know that $\nu(t)$ asymptotically converges to $\operatorname{im}(T) = \mathcal{T}$ and, within $\operatorname{im}(T)$, $\nu(t)$ persists. Therefore, the above limit is attained independently of initial conditions $\nu(0) \in \mathbb{R}^n$ if and only if

$$\lim_{t\to\infty} \left[(I \otimes (A - S)) + (C(t) \otimes BE_2)\right](1_N \otimes T) = 0.$$

Considering the diagonal structure of $C(t)$, the latter condition is equivalent to condition (3.23) given in the theorem. □

Condition (3.23) in Theorem 3.24 represents a constraint on the leader dynamics in dependence of the link strengths $c_k(t)$, $k \in \mathbb{N}_N$. There are two possibilities to satisfy condition (3.23): either $BE_2T = 0$ and $ST = AT$ in which case the limit is satisfied independently of $c_k(t)$, $k \in \mathbb{N}_N$, or the matrix equation $BXT = (S - A)T$ admits a solution $X \in \mathbb{R}^{p\times n}$. In the latter case (3.23) is satisfied if and only if $c_k(t) \to c^*$, $k \in \mathbb{N}_N$ as $t \to \infty$ for some $c^* \in \mathbb{R}_+$ and $E_2 = \frac{1}{c^*}X$. In the first case, the condition $ST = AT$ means that $\mathrm{im}(T)$ is invariant for the individual systems (3.2) with $u_k(t) \equiv 0$ and the dynamics restricted to this subspace is identical to the leader dynamics (3.20) restricted to the subspace $\mathrm{im}(T)$. That is, all modes of the leader dynamics (3.20) that are not asymptotically stable, are also modes of the individual system dynamics (3.2). In other words the persistent part of the leader model must be contained in the agent models. In the second case, $c_k(t) \to c^*$, $k \in \mathbb{N}_N$ as $t \to \infty$ means that asymptotically the leader must have direct access to all other members of the network with a constant link strength c^*.

In the case when the leader has direct access to all individuals in the network, the couplings between the individuals systems are superfluous. The problem of consensus reduces to a tracking problem for each individual system. Therefore, the more interesting case in the framework of consensus is the first case, where $ST = AT$ and $E_2 = 0$. In that case, the following result relates consensus with a leader to the consensus problems considered in Section 3.2 for fixed communication topologies:

Corollary 3.25. *Let $\mathcal{G}(t) = \{\mathcal{V}, \mathcal{E}(t), W(t)\}$ be a communication graph with $|\mathcal{V}| = N > 1$ and Laplacian matrix $L(t)$. Let $A \in \mathbb{R}^{n\times n}$, $B \in \mathbb{R}^{n\times p}$, $K \in \mathbb{R}^{p\times n}$, $S \in \mathbb{R}^{n\times n}$, $E_1 = K$, and $E_2 = 0 \in \mathbb{R}^{p\times n}$. Let $c_k(\cdot) : \mathbb{R} \to \overline{\mathbb{R}}_+$, $k \in \mathbb{N}_N$ be piecewise continuous functions of time. Let $\mathcal{T} \subset \mathbb{R}^n$ be the invariant subspace for the leader dynamics (3.20) which is maximal subject to not containing asymptotically stable modes and let $T \in \mathbb{R}^{n\times q}$ be such that $\mathrm{rank}(T) = q \in \mathbb{N}_n \bigcup\{0\}$ and $\mathrm{im}(T) = \mathcal{T}$. Define an augmented graph $\hat{\mathcal{G}}(t) = \hat{\mathcal{G}}(\hat{W}(t)) = \{\hat{\mathcal{V}}, \hat{\mathcal{E}}(\hat{W}(t)), \hat{W}(t)\}$ with $|\hat{\mathcal{V}}| = (N+1)$ vertices as the graph induced by the adjacency matrix*

$$\hat{W}(t) = \left(\begin{array}{ccc|c} & & & c_1(t) \\ & W(t) & & \vdots \\ & & & c_N(t) \\ \hline 0 & \cdots & 0 & 0 \end{array}\right) \in \mathbb{R}^{(N+1)\times(N+1)}.$$

Assume $ST = AT$.

Consensus is achieved exponentially fast in the model (3.22) over the graph $\mathcal{G}(t)$ with leader (3.20) independently of initial conditions $\nu(0) \in \mathbb{R}^n$ and $x_k(0) \in \mathbb{R}^n$, $k \in \mathbb{N}_N$ if and only if consensus is achieved exponentially fast in the model (3.4) over the extended graph $\hat{\mathcal{G}}(t)$ independently of initial conditions $x_k(0) \in \mathbb{R}^n$, $k \in \mathbb{N}_{N+1}$.

Proof. For the graph $\hat{\mathcal{G}}(t)$, the graph induced by $\{v_{N+1}\} \subset \hat{\mathcal{V}}$ is an iSCC uniformly in t by definition of $\hat{\mathcal{G}}(t)$. Thus consensus is reached exponentially if and only if $\varepsilon_k(t) \triangleq (x_k(t) - x_{N+1}(t)) \to 0$ as $t \to \infty$ exponentially fast for all $k \in \mathbb{N}_N$. We have

$$\dot{\varepsilon}(t) = [(I \otimes A) + ((L(t) + C(t)) \otimes K)]\, \varepsilon(t),$$

where $\varepsilon(t) = (\varepsilon_1^T(t), \ldots, \varepsilon_N^T(t))^T$ is the stacked vector of the errors $\varepsilon_k(t)$, $k \in \mathbb{N}_N$. On the other hand, we obtain from (3.22) with $E_1 = K$ and $E_2 = 0$

$$\dot{e}(t) = [(I \otimes A) + ((L(t) + C(t)) \otimes K)]\, e(t) + (I \otimes (A - S))(1_N \otimes \nu(t)).$$

Except for some exogenous signal $w(t) \triangleq (I \otimes (A - S))(1_N \otimes \nu(t))$, these dynamics are identical to the dynamics for $\varepsilon(t)$ derived above.

Assume $e(t) \to 0$ exponentially fast as $t \to \infty$ independent of initial conditions $e(0) \in \mathbb{R}^{Nn}$ and $\nu(0) \in \mathbb{R}^n$. Since this holds for all initial conditions, it also holds for $\nu(0) = 0$, i.e., $\nu(t) \equiv 0$. This implies that $\varepsilon(t) \to 0$ exponentially fast as $t \to \infty$ independent of initial conditions $\varepsilon(0)$.

Assume $\varepsilon(t) \to 0$ exponentially fast as $t \to \infty$ independent of initial conditions $\varepsilon(0) \in \mathbb{R}^{Nn}$. This implies that the dynamics of $\varepsilon(t)$, and thus the dynamics of $e(t)$ with input $w(t)$, are input to state stable (see Khalil, 2002). Since $AT = ST$ by Assumption, we know that $w(t) \to 0$ exponentially fast as $t \to \infty$ independent of initial conditions $\nu(0) \in \mathbb{R}$. Thus $e(t) \to 0$ exponentially fast as $t \to \infty$ independent of initial conditions $e(0) \in \mathbb{R}^{Nn}$ and $\nu(0) \in \mathbb{R}^n$. □

Together with Theorem 3.24, the above corollary has interesting implications. We need to distinguish two cases:

The first case is characterized by $AT = ST$, i.e., the persistent parts of the leader model are contained in the remaining systems' models. If this is satisfied, Corollary 3.25 suggests to choose $E_1 = K$ and $E_2 = 0$ in (3.21) and the consensus problem reduces to the consensus problems considered before, in which one system is qualified as a leader merely by specific topological properties. In particular, there is no need to distinguish between the leader and the other systems in the couplings (3.21).

The second case is characterized by $AT \neq ST$. In that case, the individual systems can only asymptotically track the leader, if they have access to the absolute value of the leader state $\nu(t)$ and it is impossible to solve the problem with diffusive couplings. It has been argued before, that this case is the less interesting case in the framework of consensus problems, since it can be reduced to tracking problems solved for the individual systems.

Thus, whenever there is a need to distinguish between the leader and the remaining systems in the coupling algorithm in order to achieve consensus, this should be considered as a hint that the problem should be solved in a tracking framework rather than in a consensus framework.

Remark 3.26. *We shortly comment on a result presented in Ren (2007) which seems to contradict our findings at first glance. Ren (2007) consider integrator agents $\dot{x}_k(t) = u_k(t)$, $k \in \mathbb{N}_N$ with a nonlinear leader modeled as $\dot{\nu}(t) = f(t, \nu(t))$. To reach consensus with the leader, the algorithm*

$$u_k(t) = \frac{\sum_{j=1}^{N} w_{k,j}\big(\dot{x}_j(t) - \gamma_k(x_k(t) - x_j(t))\big) + c_k\alpha_k\big(f(t,\nu(t)) - \gamma_k(x_k(t) - \nu(t))\big)}{\sum_{j=1}^{N} w_{k,j} + c_k\alpha_k}$$

for all $k \in \mathbb{N}_N$ is proposed, where $\alpha_k, \gamma_k \in \mathbb{R}$, $k \in \mathbb{N}_N$ are design parameters. The author claims that, in the above algorithm, only a portion of the individual systems has access to the leader state. Defining $\Gamma = \operatorname{diag}((\gamma_1, \dots, \gamma_N)^T)$ and $C_\alpha = \operatorname{diag}((c_1\alpha_1, \dots, c_N\alpha_N)^T)$ and substituting $\dot{x}_k(t) = u_k(t)$, the above algorithm can be rewritten as

$$(L + C_\alpha)(u(t) - 1_N f(t, \nu(t))) = -\Gamma(L + C_\alpha)(x(t) - 1_N\nu(t))$$

If $\operatorname{rank}(L + C_\alpha) = N$, which is claimed to be necessary and sufficient for consensus in Ren (2007), this can be solved for $u(t)$ as

$$u(t) = -(L + C_\alpha)^{-1}\Gamma(L + C_\alpha)(x(t) - 1_N\nu(t)) + 1_N f(t, \nu(t)),$$

where every local input $u_k(t)$ explicitly depends on $\nu(t)$. Hence, the result proposed in Ren (2007) is qualitatively consistent with the findings described above.

3.4.3 Connecting the Leader to an Existing Network

So far, we addressed the question whether consensus with a leader is reached over some given communication topology, i.e., we presented solutions to the State Consensus Analysis Problem 3.3 and the State Consensus Design Problem 3.4 in presence of a leader. In what follows, we slightly change our perspective. We consider a given network of systems and ask which members of the network a leader shall influence in order to make all network members asymptotically follow the leader. That is, we consider parts of the communication topology as a degree of freedom in the design. In view of the results from Section 3.4.2, we do not consider the effect of differences between the leader dynamics and the remaining systems dynamics. The result below does actually not depend on system dynamics but exclusively characterizes properties of the communication topology.

Lemma 3.27. *Let $\mathcal{G}(t) = \{\mathcal{V}, \mathcal{E}(t), W(t)\}$ be a communication graph with $|\mathcal{V}| = N > 1$. Define an augmented graph $\hat{\mathcal{G}}(t) = \hat{\mathcal{G}}(\hat{W}(t)) = \{\hat{\mathcal{V}}, \hat{\mathcal{E}}(\hat{W}(t)), \hat{W}(t)\}$ with $|\hat{\mathcal{V}}| = (N+1)$ vertices as the graph induced by the adjacency matrix*

$$\hat{W}(t) = \left(\begin{array}{ccc|c} & & & c_1(t) \\ & W(t) & & \vdots \\ & & & c_N(t) \\ \hline 0 & \cdots & 0 & 0 \end{array}\right) \in \mathbb{R}^{(N+1)\times(N+1)}.$$

Assume the graph $\mathcal{G}(t)$ contains exactly r distinct iSCCs $\tilde{\mathcal{G}}_k(t) = \{\tilde{\mathcal{V}}_k, \tilde{\mathcal{E}}_k(t), \tilde{W}_k(t)\}$ with $k \in \mathbb{N}_r$ uniformly in t.

Then $\hat{\mathcal{G}}(t)$ is uniformly connected if and only if there exist a threshold $\alpha \in \mathbb{R}_+$, a time horizon $T \in \mathbb{R}_+$, and integers $l_k \in \mathbb{N}_N$ such that $v_{l_k} \in \tilde{\mathcal{V}}_k$ for $k \in \mathbb{N}_r$ and $\int_t^{t+T} c_{l_k}(\tau)\mathrm{d}\tau > \alpha$, $k \in \mathbb{N}_r$ uniformly in t.

Proof. Let $\overline{W}_T(t) \triangleq \int_t^{t+T} \hat{W}(\tau)\mathrm{d}\tau$ and $\overline{\mathcal{G}}_T(t) \triangleq \mathcal{G}(\overline{W}_T(t)) = \{\hat{\mathcal{V}}, \mathcal{E}(\overline{W}_T(t)), \overline{W}_T(t)\}$. By definition of $\hat{\mathcal{G}}(t)$, the subgraph induced by $\{v_{N+1}\} \subset \hat{\mathcal{V}}$ is an iSCC of $\hat{\mathcal{G}}(t)$ uniformly in t. Therefore, the subgraph of $\overline{\mathcal{G}}_T(t)$ induced by $\{v_{N+1}\} \subset \hat{\mathcal{V}}$ is also an iSCC of $\overline{\mathcal{G}}_T(t)$ uniformly in t and $\hat{\mathcal{G}}(t)$ is uniformly connected if and only if there exists some $T \in \mathbb{R}_+$ such that the subgraph induced by $\{v_{N+1}\} \subset \hat{\mathcal{V}}$ is the unique iSCC of $\overline{\mathcal{G}}_T(t)$ uniformly in t.

Assume $\alpha \in \mathbb{R}^+$, $T \in \mathbb{R}^+$, and $l_k \in \mathbb{N}_N$, $k \in \mathbb{N}_r$ are such that $v_{l_k} \in \tilde{\mathcal{V}}_k$ for $k \in \mathbb{N}_r$ and $\int_t^{t+T} c_{l_k}(\tau)\mathrm{d}\tau > \alpha$, $k \in \mathbb{N}_r$ uniformly in t. Then $(v_{N+1}, v_{l_k}) \in \mathcal{E}(\overline{W}_T(t))$, $k \in \mathbb{N}_r$. Since for any $t \in \mathbb{R}$ and for every vertex $v \in \mathcal{V}$ there exists a vertex $w \in \bigcup_{k \in \mathbb{N}_r} \tilde{\mathcal{V}}_k$ such that there exists a path from w to v in $\mathcal{G}(t)$ by Theorem 2.12. Thus, there also exists a path from v_{N+1} to any vertex $v \in \mathcal{V}$ at any time $t \in \mathbb{R}$ in the union graph $\overline{\mathcal{G}}_T(t)$. This shows that the subgraph induced by $\{v_{N+1}\} \subset \hat{\mathcal{V}}$ is the unique iSCC of $\overline{\mathcal{G}}_T(t)$.

Assume on the other hand that the subgraph induced by $\{v_{N+1}\} \subset \hat{\mathcal{V}}$ is the unique iSCC of $\overline{\mathcal{G}}_T(t)$. Then there exists a path from v_{N+1} to any vertex $v \in \mathcal{V}$, in particular to those vertices belonging to the iSCCs $\tilde{\mathcal{G}}_k(t)$, $k \in \mathbb{N}_r$. This implies that there exist integers

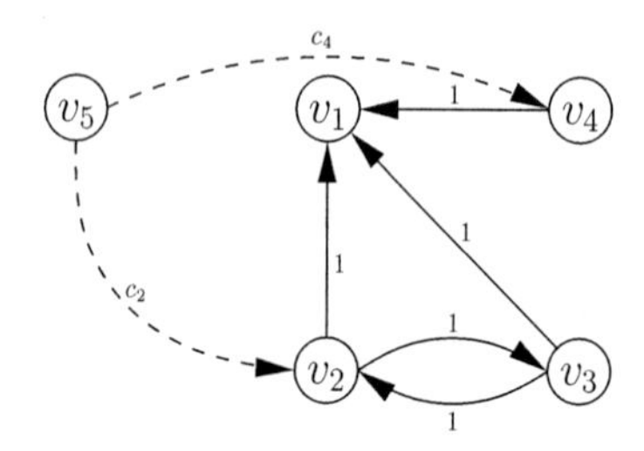

Figure 3.5: Graph with four vertices with two iSCCs (solid lines) augmented by a leader v_5 in a way to obtain a connected graph (dashed lines).

$l_k \in \mathbb{N}_N$, $k \in \mathbb{N}_r$ such that $v_{l_k} \in \tilde{\mathcal{V}}_k$ and $(v_{N+1}, v_{l_k}) \in \overline{\mathcal{E}}_T(t)$ for all $k \in \mathbb{N}_r$. The latter statement in turn implies existence of a threshold $\alpha \in \mathbb{R}_+$ and a time horizon $T \in \mathbb{R}_+$ such that $\int_t^{t+T} c_{l_k}(\tau)\mathrm{d}\tau > \alpha$, $k \in \mathbb{N}_r$ uniformly in t. □

Lemma 3.27 tells us, that in order to influence a network, it suffices to influence just one system for every iSCC contained in the network. In case of constant communication graphs, we can use Theorem 2.13 to identify the links between the leader and the existing network needed to enable consensus as follows:

(S1) Construct an orthogonal non-negative basis of $\ker(L^T)$. Let $r = \dim(\ker(L^T))$ and denote the basis vectors p_k, $k \in \mathbb{N}_r$.

(S2) Connect the leader to r vertices v_{l_k} such that $p_{k,l_k} > 0$, $k \in \mathbb{N}_r$.

Each basis vector constructed in Step (S1) corresponds to one iSCC of the graph. Step (S2) thus ensures, that the leader is connected to each iSCC in the graph, i.e., the leader is the unique iSCC of the graph augmented by one vertex corresponding to the leader. An example is depicted in Figure 3.5. The original graph (solid lines) contains two iSCCs, namely the graphs induced by $\{v_4\}$ and $\{v_2, v_3\}$. Accordingly, the kernel of L^T is spanned by the vectors $p_1 = (0,0,0,1)^T$ and $p_2 = (0,1,1,0)^T$. Connecting the leader with as depicted in Figure 3.5 (dashed lines) yields a connected graph.

3.5 Summary

In this chapter, we investigated several instances of consensus problems employing static diffusive couplings. We posed two basic problems, namely the State Consensus Analysis Problem 3.3 and the State Consensus Design Problem 3.4 and presented solutions to those problems under different assumptions and preconditions.

In a first part, necessary and sufficient conditions for consensus of general LTI systems over fixed communication topologies have been presented and an LMI based design procedure was developed, based on Wieland et al. (2008, 2010a). These results include an improvement on existing bounds on the coupling gains for consensus among double integrator systems.

The second part has been devoted to consensus of LTI systems over communication topologies that may change over time. Results known from Literature have been reviewed and compared to the results for fixed communication topologies.

It has been demonstrated that there is a trade-off between allowable topological complexity and system complexity in consensus via static diffusive couplings. While consensus can be achieved among arbitrary stabilizable LTI systems over constant communication graphs, consensus over uniformly communication graphs requires system models that do not contain exponentially unstable parts and, more importantly, where the input dimension equals the state dimension. This observation motivates the quest for dynamic couplings that allow to move beyond the compromise between system and topological complexity we are forced to accept in case of static couplings and to further relax constraints on individual systems as well as the communication topology.

Finally, we considered the special case of consensus with a leader and showed based on Wieland et al. (2008, 2010a) that the leader dynamics is constrained to be essentially identical to the dynamics of the remaining systems in the network in meaningful consensus problems. Furthermore, we proposed an algorithm to determine how to connect a leader to an existing network in order to achieve consensus.

Chapter 4

Observer-Based Output Consensus with Relative Output Sensing

In Chapter 3, the problem of state consensus by means of static diffusive couplings has been considered. The different solutions proposed for the various instances of problems that have been investigated all share one common feature. Namely, consensus is achieved by relative *state feedback*, i.e., in order to achieve consensus, information of the relative values of the full states are required. Compared to the feedback stabilization problem, the situation corresponds to stabilization by state feedback.

In practice, the complete state vector is often not available for measurement. Instead, the only information available for measurement is frequently some output of dimension much lower than the complete state vector. Thus, feedback control problems have to be solved based on the only information of the measured output. In many instances, it is possible to design dynamic observers that yield estimates of the original state vector and to apply a state feedback in which the actual state vector is replaced by its estimate. The well-known separation principle (Kailath, 1980) gives a sufficient condition for this approach to be valid and to actually solve the original control problem.

In this chapter, we investigate to what extent these ideas can be adopted to consensus problems. Similarly to the case of feedback stabilization, it might be unrealistic in consensus problems to assume complete knowledge of the relative states. It is rather the case, that only relative outputs are available to solve the consensus problem. These relative outputs stem, e.g., from relative sensors. For instance in a group of mobile vehicles, sensors may be available that measure relative positions or relative velocities. In what follows, we will thus tackle the question, how observer design techniques commonly used in feedback stabilization problems can be adapted to the consensus problem.

The problem of consensus with relative output sensing is non-trivial. In standard observer designs for linear systems, information on the output and *the input* is required in order to reconstruct the system state. While this is frequently not a restriction in feedback stabilization problems, the situation is more difficult in consensus problems with relative output sensing. In that case information on *relative* inputs in required. Yet, this information depends on the inputs of the neighboring systems and is thus unknown. Subsequently, the difficulties that occur in this context will be detailed and solutions to overcome these problems will be proposed. Time-varying communication topologies may induce additional difficulties for observer designs in the context of output consensus problems, but we will restrict our attention to challenges induced by relative output sensing over constant communication topologies.

Since the objective of this chapter is to design dynamic observers to be used in the couplings between individual systems, we will for the first time in this thesis fulfill the promise to demonstrate the benefit of dynamic couplings in consensus and synchronization problems. As argued before, in this chapter, we use dynamic couplings to overcome the specific constraint of full state information in order to achieve consensus, i.e., to relax conditions on the individual system dynamics.

4.1 Problem Statement

In contrast to Chapter 3, where LTI systems (3.2) without explicit output were considered, we now consider networks of N individual systems modeled as

$$\dot{x}_k(t) = Ax_k(t) + Bu_k(t) \tag{4.1a}$$
$$y_k(t) = Cx_k(t) \tag{4.1b}$$

for $k \in \mathbb{N}_N$, where $x_k(t) \in \mathbb{R}^n$ is the state vector, $u_k(t) \in \mathbb{R}^p$ is the input vector and $y_k(t) \in \mathbb{R}^q$ is the output vector. Without loss of generality, we assume that B has full column-rank and C has full row-rank, i.e., $\operatorname{rank}(B) = p$ and $\operatorname{rank}(C) = q$.

4.1.1 Dynamic Diffusive Couplings and Output Consensus Trajectories

We are no longer interested in state consensus in the sense of Definition 3.1 but rather consider output consensus as the property that $(y_k(t) - y_j(t)) \to 0$ as $t \to \infty$ for all $j, k \in \mathbb{N}_N$. We give the formal counterpart to Definition 3.1 below.

Definition 4.1 (Output Consensus). *The N systems (4.1) are said to asymptotically reach output consensus, if there exists some trajectory $\eta(\cdot) : \mathbb{R} \to \mathbb{R}^q$ such that*

$$\lim_{t\to\infty} \|y_k(t) - \eta(t)\| = 0 \tag{4.2}$$

for all $k \in \mathbb{N}_N$. In that case, the trajectory $\eta(\cdot)$ is called an output consensus trajectory.

The static couplings (3.3) from Chapter 3 are replaced by dynamic couplings given as

$$\dot{z}_k(t) = Ez_k(t) + F\delta_k(t) \tag{4.3a}$$
$$u_k(t) = Gz_k(t) + H\delta_k(t) \tag{4.3b}$$
$$\delta_k(t) = \sum_{j=1}^{N} w_{k,j}(t)(y_k(t) - y_j(t)) \tag{4.3c}$$

with coupling state $z_k(t) \in \mathbb{R}^m$ for $k \in \mathbb{N}_N$. Couplings (4.3) represent a general LTI dynamic controller driven by the relative output signal $\delta_k(t)$ as defined by the last equation (4.3c). Similarly to (3.3), the values $w_{k,j}(t) \in \overline{\mathbb{R}}_+$, $j, k \in \mathbb{N}_N$ are the elements of the adjacency matrix $W(t) = [w_{k,j}(t)] \in \mathbb{R}^{N\times N}$ of some communication graph $\mathcal{G}(t) = \{\mathcal{V}, \mathcal{E}(t), W(t)\}$ with $|\mathcal{V}| = N$. We put some constraints on the coupling matrices $E \in \mathbb{R}^{m\times m}$, $F \in \mathbb{R}^{m\times q}$, $G \in \mathbb{R}^{p\times m}$, and $H \in \mathbb{R}^{p\times q}$ formulated as an assumption below.

Assumption 4.2. *The matrix $E \in \mathbb{R}^{m\times m}$ is Hurwitz, the matrix $F \in \mathbb{R}^{m\times q}$ is such that the pair (E, F) is controllable, and the matrix $G \in \mathbb{R}^{p\times m}$ is such that the pair (G, E) is observable.*

Assuming controllability and observability in (4.3a), (4.3b) is in fact not a restriction. It essentially means that we choose a minimal realization (see Kailath, 1980) of the dynamic couplings. Requiring E to be Hurwitz means that the coupling dynamics are asymptotically – and thus exponentially – stable. Therefore, $\delta_k \to 0$ as $t \to \infty$ exponentially fast implies by Lemma B.1 that $u_k(t) \to 0$ as $t \to \infty$ exponentially fast. With this last property couplings (4.3) can be considered a generalization of static diffusive couplings as defined in Section 1.2 to dynamic diffusive couplings. While the system input $u_k(t)$ is no longer proportional to a weighted average of relative variables uniformly in t, this relation still holds when we consider steady states, i.e., in the limit as $t \to \infty$, if the limits exist.

In case of static diffusive couplings we showed in Lemma 3.2 that consensus trajectories are open loop solutions of the individual system dynamics if convergence to consensus is exponential. Using Assumption 4.2, we are able to give an analogous result for output consensus in the network (4.1), (4.3).

Lemma 4.3. *Let $\tilde{\eta}(\cdot) : \mathbb{R} \to \mathbb{R}^q$ be an output consensus trajectory for* (4.1), (4.3) *for fixed initial conditions $x_k(0) \in \mathbb{R}^n$ and $z_k(0) \in \mathbb{R}^m$, $k \in \mathbb{N}_N$.*

If (C, A) is observable, the matrices E, F, and G in (4.3) *satisfy Assumption 4.2, and* (4.2) *is satisfied with an exponential convergence rate, i.e., there exist constants $M_1, \mu_1 \in \mathbb{R}_+$ such that*

$$\|y_k(t) - \tilde{\eta}(t)\| \leq M_1 e^{-\mu_1 t} \max_{j\in\mathbb{N}_N} \|y_j(0) - \tilde{\eta}(0)\|$$

for all $k \in \mathbb{N}_N$, then there exists $\xi_0 \in \mathbb{R}^n$ and constants $M_2, \mu_2 \in \mathbb{R}_+$ such that

$$\|y_k(t) - Ce^{At}\xi_0\| \leq M_2 e^{-\mu_2 t} \max_{j\in\mathbb{N}_N} \|y_j(0) - C\xi_0\|$$

for all $k \in \mathbb{N}_N$, i.e., the output $\eta(t) = Ce^{At}\xi_0$ of the open loop dynamics (4.1) *with $u_k(t) \equiv 0$ initialized as $x_k(0) = \xi_0$ is an output consensus trajectory for* (4.1), (4.3) *with exponential convergence rate for the same initial conditions $x_k(0) \in \mathbb{R}^n$ and $z_k(0) \in \mathbb{R}^m$, $k \in \mathbb{N}_N$.*

Proof. If $\tilde{\eta}(t)$ is a consensus trajectory, then $y(t)$ asymptotically converges to the subspace $\mathrm{im}(1_N \otimes I_q) \subset \mathbb{R}^{Nq}$. By observability of the pair (C, A), it follows that $x(t)$ asymptotically converges to the subspace $\mathrm{im}(1_N \otimes I_n) \subset \mathbb{R}^{Nn}$. If $x(t) \in \mathrm{im}(1_N \otimes I_n)$, then $x(t) = 1_N \otimes \xi(t)$ and $y(t) = 1_N \otimes (C\xi(t))$ for some trajectory $\xi(\cdot) : \mathbb{R} \to \mathbb{R}^n$. Therefore $\delta(t) = (L(t) \otimes C)x(t) = (L(t) \otimes I_q)y(t) = 0$, i.e., $(x(t), z(t))$ converges to $\mathrm{im}(1_N \otimes I_n) \times \{0\}$ and $\mathrm{im}(1_N \otimes I_n) \times \{0\} \subset \mathbb{R}^{Nn} \times \mathbb{R}^{Nm}$ is an invariant subspace for the closed loop system.

Convergence of $(x(t), z(t))$ to the subspace $\mathrm{im}(1_N \otimes I_n) \times \{0\} \subset \mathbb{R}^{Nn} \times \mathbb{R}^{Nm}$ is exponential by assumption. In fact $\delta(t) \to 0$ exponentially fast as $t \to \infty$ implies that $x(t)$ converges to $\mathrm{im}(1_N \otimes I_n)$ exponentially fast as $t \to \infty$ by observability of the pair (C, A) and $z(t) \to 0$ exponentially fast as $t \to \infty$ by Lemma B.1. Using again Lemma B.1, this implies that $(x(t), z(t))$ exponentially converges to a particular solution $(1_N \otimes \xi(t)) \times 0$ contained in $\mathrm{im}(1_N \otimes I_n) \times \{0\}$. This completes the proof. □

Similarly to Lemma 3.2 in the state consensus case, Lemma 4.3 relates output consensus trajectories to solutions of the open loop system (4.1) with $u_k(t) \equiv 0$. The key property of the couplings (4.3) used in the above proof is that the couplings *exponentially* vanish once consensus is reached. In addition, we assume observability of (C, A) in Lemma 4.3. This assumption can be dropped by decomposing the system in observable and unobservable subsystems (see Kailath, 1980) and applying the above lemma to the observable subsystem only.

4.1.2 Limitations of Luenberger Observers in Output Consensus with Relative Output Sensing

In view of the results of Chapter 3, the most proximate approach to design dynamic couplings (4.3) might seem to design Luenberger observers (Kailath, 1980, Chapter 4) to obtain asymptotic estimates for the relevant states and use the methods from Chapter 3 where true system states are replaced by their estimates.

In order to realize couplings (3.3) with the help of asymptotic observers, we need to estimate the weighted sum of relative states $\rho_k(t) \triangleq \sum_{j=1}^{N} w_{k,j}(t)(x_k(t) - x_j(t))$ appearing in the couplings (3.3). With $\gamma_k(t) \triangleq \sum_{j=1}^{N} w_{k,j}(t)(u_k(t) - u_j(t))$ and (4.3c), these relative state vectors evolve according to the dynamics

$$\begin{aligned}\dot{\rho}_k(t) &= A\rho_k(t) + B\gamma_k(t) + \sum_{j=1}^{N} \dot{w}_{k,j}(t)(x_k(t) - x_j(t)) \\ \delta_k(t) &= C\rho_k(t)\end{aligned}$$

for all $k \in \mathbb{N}_N$. If the pair (C, A) is detectable, classical Luenberger observers for these systems are given as

$$\dot{\hat{\rho}}_k(t) = (A + JC)\hat{\rho}_k(t) - J\delta_k(t) + B\gamma_k(t) + \sum_{j=1}^{N} \dot{w}_{k,j}(t)(x_k(t) - x_j(t))$$

for all $k \in \mathbb{N}_N$, where $\hat{\rho}_k(t) \in \mathbb{R}^n$ is an estimate for $\rho_k(t) \in \mathbb{R}^n$ for $k \in \mathbb{N}_N$ and $J \in \mathbb{R}^{n \times q}$ is the observer gain chosen such that $A + JC$ is Hurwitz. Obviously, these observer dynamics structurally differ from the dynamics (4.3a) of the couplings we aim to design. The above observer dynamics depend on the relative inputs $\gamma_k(t)$, $k \in \mathbb{N}_N$ and the sum $\sum_{j=1}^{N} \dot{w}_{k,j}(t)(x_k(t) - x_j(t))$, $k \in \mathbb{N}_N$, both of which are unknown. The difficulties induced by the latter term are manifold: it cannot be measured, it may take any value in $\mathbb{R}^n$, i.e., if considered as an input, the corresponding input matrix would be the identity matrix, and the input signal might exhibit impulsive behavior at time instances $t \in \mathbb{R}$ where $w_{k,j}(t), j \in \mathbb{N}_N$ is not continuous. We avoid those problems by restricting the analysis to problems with constant communication topologies in what follows. We instead focus on the difficulty induced by the unknown relative inputs $\gamma_k(t)$, $k \in \mathbb{N}_N$. That is, we aim at observers for the systems

$$\dot{\rho}_k(t) = A\rho_k(t) + B\gamma_k(t) \tag{4.4a}$$

$$\delta_k(t) = C\rho_k(t) \tag{4.4b}$$

for $k \in \mathbb{N}_N$, that yield asymptotic estimates $\hat{\rho}_k(t) \in \mathbb{R}^n$ for $\rho_k(t)$ based on output measures $\delta_k(t) \in \mathbb{R}^q$ with unknown input $\gamma_k(t) \in \mathbb{R}^p$. These observers shall be such that state consensus for the systems (4.1a) is achieved with $u_k(t) = K\rho_k(t)$ (corresponding to the couplings (3.3) in Chapter 3) if and only if output consensus is reached for the systems (4.1) with $u_k(t) = K\hat{\rho}_k(t)$.

One approach to solve this problem is to take advantage of the fact that we are not interested in the actual value of the estimate $\hat{\rho}_k(t)$, but rather want to use it for a specific task, namely to achieve consensus. Fulfillment of this task implies $\gamma_k(t) \to 0$. Thus, we actually do not need to design an observer, that yields a perfect estimate for any input $\gamma_k(t)$ but it is enough if an imperfect observer is used that ensures $\gamma_k(t) \to 0$ when combined with an appropriate static feedback. This idea is pursued in Kim et al. (2010), Seo et al. (2009a,b). However, since this observer only delivers and asymptotic estimate of the true system state under the condition that $\gamma_k(t) \to 0$, the separation principle no longer holds. Consequently the observer design and the static feedback design are no longer independent, which makes it difficult to apply the results from the previous chapter.

Therefore, we pursue a different approach. Namely, we aim at designing general observers that yield state estimates independent of the input, i.e., *unknown-input observers*. The advantage of this approach is the possibility to design observers and state feedback independent of each other, i.e., the separation principle holds. Furthermore, the use of unknown-input observers yields nice robustness properties. The results below are based on Wieland and Allgöwer (2010).

4.2 Consensus with Full Order Unknown-Input Observers

Our first approach is to use unknown input observers as proposed in Darouach et al. (1994). The observers for the systems (4.4) take the general form

$$\dot{z}_k(t) = Ez_k(t) + F\delta_k(t) \tag{4.5a}$$
$$\hat{\rho}_k(t) = Rz_k(t) + S\delta_k(t) \tag{4.5b}$$

for $k \in \mathbb{N}_N$, with observer state $z_k \in \mathbb{R}^m$, $m \leq n$, and estimate $\hat{\rho}_k(t) \in \mathbb{R}^n$ for $k \in \mathbb{N}_N$. Necessary and sufficient existence conditions for such an observer are stated in the following lemma taken from Darouach et al. (1994) and repeated here for reference and without proof:

Lemma 4.4. *Let $A \in \mathbb{R}^{n\times n}$, $B \in \mathbb{R}^{n\times p}$, and $C \in \mathbb{R}^{q\times n}$.*

Assume $\operatorname{rank}(B) = p$ *and* $\operatorname{rank}(C) = q$.

There exist matrices $E \in \mathbb{R}^{m\times m}$, $F \in \mathbb{R}^{m\times q}$, $R \in \mathbb{R}^{n\times m}$, and $S \in \mathbb{R}^{n\times q}$ such that (4.5) is an asymptotic observer for (4.4), i.e., $\rho_k(t) - \hat{\rho}_k(t) \to 0$ as $t \to \infty$ for all initial conditions $\rho_k(0) \in \mathbb{R}^n$, $z_k(0) \in \mathbb{R}^m$ if and only if

$$\operatorname{rank}(CB) = \operatorname{rank}(B) = p \tag{4.6}$$

and

$$\operatorname{rank}\begin{pmatrix} A - sI_n & B \\ C & 0 \end{pmatrix} = n + p, \quad s \in \overline{\mathbb{C}}_+. \tag{4.7}$$

The conditions given in the above lemma have direct systems theoretic interpretations. Namely, condition (4.6) implies that system (4.4) has relative degree one with respect to every scalar output while condition (4.7) is equivalent to requiring asymptotically stable zero dynamics. If $p = q$, i.e., system (4.4) has the same number of inputs and outputs, the pair (A, B) is controllable, and the pair (C, A) is observable, conditions (4.6), (4.7) are satisfied if and only if system (4.4) is feedback equivalent to a strictly passive system (Byrnes et al., 1991, Sepulchre et al., 1997).

Given the matrices $A \in \mathbb{R}^{n\times n}$, $B \in \mathbb{R}^{n\times p}$ and $C \in \mathbb{R}^{q\times n}$ satisfy the conditions of Lemma 4.4, the following result gives a method to design the dynamic couplings (4.3) ensuring output consensus among the systems (4.1):

Theorem 4.5. *Let $\mathcal{G} = \{\mathcal{V}, \mathcal{E}, W\}$ be a constant communication graph with $|\mathcal{V}| = N > 1$ and Laplacian matrix L. Let $A \in \mathbb{R}^{n\times n}$, $B \in \mathbb{R}^{n\times p}$, $C \in \mathbb{R}^{q\times n}$, and $K \in \mathbb{R}^{p\times n}$.*

Assume $\mathcal{G}$, A, B, and K satisfy the assumptions and conditions of Theorem 3.7 and A, B, and C satisfy the assumptions and conditions of Lemma 4.4.

Then it is possible to choose $J \in \mathbb{R}^{n\times q}$ such that $PA + JC$ is Hurwitz, where $P \triangleq I_n - SC$, $S \triangleq B(CB)^+$.

In that case, the dynamic couplings (4.3) with $m = n$, $E = PA + JC$, $F = PAS - J(I_q - CS)$, $G = K$, and $H = KS$ ensure that systems (4.1) reach consensus exponentially fast independent of initial conditions $x_k(0) \in \mathbb{R}^n$, $z_k(0) \in \mathbb{R}^n$, $k \in \mathbb{N}_N$.

Proof. We first consider the observer (4.5) with $R = I_n$. The observer error is $\varepsilon_k(t) = \rho_k(t) - \hat{\rho}_k(t)$ and satisfies

$$\begin{aligned}
\dot{\varepsilon}_k(t) &= (I_n - SC)(A\rho_k(t) + B\gamma_k(t)) - Ez_k(t) - F\delta_k(t)\\
&= PA\rho_k(t) - E(\hat{\rho}_k(t) - S\delta_k(t)) - F\delta_k(t)\\
&= (PA + JC)\varepsilon_k(t) - (J - (PA + JC)S + PAS - J(I_q - CS))\delta_k(t)\\
&= (PA + JC)\varepsilon_k(t) = E\varepsilon_k(t),
\end{aligned}$$

where we used $PB = (I_n - B(CB)^+C)B = 0$ as a consequence of condition (4.6). Thus, if J is chosen such that $E = PA + JC$ is Hurwitz, the estimate $\hat{\rho}_k(t)$ exponentially converges to the true system state $\rho_k(t)$.

With $\hat{\rho}_k(t) = \rho_k(t) - \varepsilon_k(t)$ and $u_k(t) = K\hat{\rho}_k(t) = K\sum_{j=1}^{N} w_{k,j}(x_k(t) - x_j(t)) - K\varepsilon_k(t)$, $k \in \mathbb{N}_N$, it follows that the closed loop dynamics for the whole network is given as

$$\begin{pmatrix} \dot{x}(t) \\ \dot{\varepsilon}(t) \end{pmatrix} = \begin{pmatrix} (I_N \otimes A) + (L \otimes BK) & -(I_N \otimes BK) \\ 0 & (I_N \otimes E) \end{pmatrix} \begin{pmatrix} x(t) \\ \varepsilon(t) \end{pmatrix}$$

with stacked vectors $x(t) = (x_1^T(t), \ldots, x_N^T(t))^T \in \mathbb{R}^{Nn}$ and $\varepsilon(t) = (\varepsilon_1^T(t), \ldots, \varepsilon_N^T(t))^T \in \mathbb{R}^{Nn}$. That is, the separation principle indeed holds.

It remains to show that the given assumptions are sufficient to guarantee existence of some matrix $J \in \mathbb{R}^{n\times q}$ such that $PA + JC$ is Hurwitz, which is equivalent to detectability of the pair (C, PA) (see Kailath, 1980). However, detectability of the pair (C, PA) is equivalent to condition (4.7) by Darouach et al. (1994, Theorem 2). This completes the proof. $\square$

Lemma 4.4 provides necessary and sufficient conditions for existence of an unknown-input observer to estimate the quantities $\rho_k(t) = \sum_{j=1}^{N} w_{k,j}(x_k(t) - x_j(t))$ and Theorem 4.5

provides an observer based design method to obtain dynamic couplings of the type (4.3) that yield consensus in the network (4.1), (4.3) with coupling gain K determined using the methods from Chapter 3. Unfortunately, the existence conditions for the unknown-input observer (4.5) specified in Lemma 4.4 are rather restrictive. In the next section we will try to modify this design to obtain less restrictive existence conditions.

4.3 Consensus with Reduced Order Unknown-Input Observers

The observers in the solution proposed in Theorem 4.5 yield estimates $\hat{\rho}_k(t) \in \mathbb{R}^n$, $k \in \mathbb{N}_N$ for the full states $\rho_k(t) \in \mathbb{R}^n$, $k \in \mathbb{N}_N$. However, to solve the output consensus problem, these estimates are never used directly, but only the quantities $K\hat{\rho}_k(t)$ are needed to achieve output consensus. Therefore, in what follows, we investigate to what extent the existence conditions (4.6), (4.7) can be relaxed if we aim for estimates of the quantities $K\rho_k(t)$ directly, using reduced order unknown-input observers.

The approach used to obtain these estimates is the following: In a first step, we address the question which parts of the states $\rho_k(t)$ of systems (4.4) can be asymptotically estimated without knowledge of the inputs $\gamma_k(t)$ without assuming (4.6) and (4.7). That is, we seek maximal unknown input observers for a given system. In a second part, we then try to find couplings with methods from Chapter 3. The methods will be adapted such that the couplings depend only on those parts of the state vectors $\rho_k(t)$ for which an asymptotic estimate is available.

4.3.1 Maximal Functional Unknown-Input Observers

To estimate parts of the states $\rho_k(t) \in \mathbb{R}^n$ of systems (4.4), we use unknown input observers

$$\dot{z}_k(t) = Ez_k(t) + F\delta_k(t) \tag{4.8a}$$

$$\hat{\zeta}_k(t) = Rz_k(t) + S\delta_k(t) \tag{4.8b}$$

for $k \in \mathbb{N}_N$, with state vectors $z_k(t) \in \mathbb{R}^m$ and estimate $\hat{\zeta}_k(t) \in \mathbb{R}^r$, $r \leq n$, such that $\hat{\zeta}_k(t)$ is an asymptotic estimate for $\zeta_k(t) = D\rho_k(t) \in \mathbb{R}^r$ independent of initial conditions $\rho_k(0) \in \mathbb{R}^n$, $z_k(0) \in \mathbb{R}^m$ for some matrix $D \in \mathbb{R}^{r\times n}$ and $\rho_k(t)$ evolving according to (4.4). We use the following definition subsequently:

Definition 4.6 (Unknown-Input Observable, Complementary Unknown-Input Observable Subspace). *Let $A \in \mathbb{R}^{n\times n}$, $B \in \mathbb{R}^{n\times p}$, $C \in \mathbb{R}^{q\times n}$, and $D \in \mathbb{R}^{r\times n}$.*

If there exist matrices $E \in \mathbb{R}^{m\times m}$, $F \in \mathbb{R}^{m\times q}$, $R \in \mathbb{R}^{r\times m}$, and $S \in \mathbb{R}^{r\times q}$ such that the observer (4.8) yields an asymptotic estimate $\hat{\zeta}_k(t)$ for $\zeta_k(t) = D\rho_k(t)$ independent of initial conditions $\rho_k(0) \in \mathbb{R}^n$, $z_k(0) \in \mathbb{R}^m$, where $\rho_k(t)$ evolves according to the dynamics (4.4), then $\zeta_k(t) = D\rho_k(t)$ is said to be unknown-input observable *and* $\ker(D) \subset \mathbb{R}^n$ *is the corresponding* complementary unknown-input observable subspace.

We are interested in recovering as much information as possible about the original system state through the observer (4.8). Therefore, we are interested in the maximal unknown input observer of the type (4.8) in the following sense:

Lemma 4.7. *Let $A \in \mathbb{R}^{n \times n}$, $B \in \mathbb{R}^{n \times p}$, and $C \in \mathbb{R}^{q \times n}$. Let $\mathbb{T}$ be the set of all complementary unknown-input observable subspaces for system* (4.4)

Then there exists a complementary unknown-input observable subspace $\mathcal{T}_{\min} \in \mathbb{T}$ such that $\mathcal{T}_{\min} \subseteq \mathcal{T}$ for any $\mathcal{T} \in \mathbb{T}$.

Proof. Assume $\mathcal{T}_1, \mathcal{T}_2 \in \mathbb{T}$ with $\dim(\mathcal{T}_1) = n - r_1$ and $\dim(\mathcal{T}_2) = n - r_2$. Choose $D_1 \in \mathbb{R}^{r_1 \times n}$ and $D_2 \in \mathbb{R}^{r_2 \times n}$ such that $\ker(D_1) = \mathcal{T}_1$ and $\ker(D_2) = \mathcal{T}_2$, i.e., $D_1 \rho_k(t)$ and $D_2 \rho_k(t)$ are unknown-input observable. Consequently, $(D_1^T, D_2^T)^T \rho_k(t)$ is also unknown-input observable and $\ker((D_1^T, D_2^T)^T) = \mathcal{T}_1 \bigcap \mathcal{T}_2 \in \mathbb{T}$. That is, $\mathbb{T}$ is closed under subspace intersection. Since $\mathbb{R}^n$ is finite dimensional and $\ker(C) \in \mathbb{T}$, there exists an infimal member $\mathcal{T}_{\min} \in \mathbb{T}$. □

The minimal complementary unknown-input observable subspace $\mathcal{T}_{\min}$ from the above lemma yields the maximal unknown-input observable part of the state $\rho_k(t)$ as follows. Let $r = n - \dim(\mathcal{T}_{\min}) \geq q$ and choose any $D \in \mathbb{R}^{r \times n}$ such that $\ker(D) = \mathcal{T}_{\min}$. Then $\zeta_k(t) = D\rho_k(t)$ is maximal subject to being unknown-input observable and unique modulo linear transformations. In what follows, we explain how to determine $\mathcal{T}_{\min}$ and how to choose the observer matrices in (4.8) to estimate the maximal unknown input-observable $\zeta_k(t)$.

In order to construct the observer, we need some geometric concepts from linear control theory summarized in Appendix C. As a consequence of Lemma C.4, we know that $\mathcal{T}_{\min} = \mathcal{S}^*_{g,\mathrm{im}(B)} \bigcup \ker(C)$, where $\mathcal{S}^*_{g,\mathrm{im}(B)}$ is the minimal complementary detectability subspace containing $\mathrm{im}(B)$.

The procedure to construct the maximal unknown-input observer is summarized in the following algorithm:

Algorithm 4.8.

(S1) Determine the conditionally invariant subspace (see Definition C.1) $\mathcal{S}^*_{\mathrm{im}(B)}$ which is minimal subject to containing $\mathrm{im}(B)$ *using Algorithm C.5.*

(S2) Determine the complementary observability subspace (see Definition C.3) $\mathcal{M}^*_{\mathrm{im}(B)}$ which is minimal subject to containing $\mathrm{im}(B)$ *using Algorithm C.6.*

(S3) Choose $\chi \in \mathbb{R}_+$ and define $\mathbb{C}_g \triangleq \{z \in \mathbb{R} | \mathfrak{Re}(z) < -\chi\}$ and $\mathbb{C}_b \triangleq \mathbb{C} \setminus \mathbb{C}_g$.

(S4) Choose any friend (see Definition C.1) $J_0 \in \boldsymbol{J}(\mathcal{S}^*_{\mathrm{im}(B)})$. Determine $\mathcal{X}_g \subset \mathbb{R}^n$ and $\mathcal{X}_b \subset \mathbb{R}^n$ such that $(A+J_0C)\mathcal{X}_g \subset \mathcal{X}_g$ and $(A+J_0C)\mathcal{X}_b \subset \mathcal{X}_b$, $\sigma((A+J_0C)|\mathcal{X}_g) \subset \mathbb{C}_g$ and $\sigma((A+J_0C)|\mathcal{X}_b) \subset \mathbb{C}_b$, and $\mathcal{X}_g \oplus \mathcal{X}_b = \mathbb{R}^n$. Then $\mathcal{S}^*_{g,\mathrm{im}(B)}$ is obtained as

$$\mathcal{S}^*_{g,\mathrm{im}(B)} = \mathcal{S}^*_{\mathrm{im}(B)} + \mathcal{X}_b \bigcap \mathcal{M}^*_{\mathrm{im}(B)}.$$

Define $s \triangleq n - \dim\left(\mathcal{S}^*_{g,\mathrm{im}(B)}\right)$ and $r \triangleq n - \dim\left(\mathcal{S}^*_{g,\mathrm{im}(B)} \bigcap \ker(C)\right)$.

(S5) Let $P \in \mathbb{R}^{s \times n}$ be a matrix representation of the canonical projection $P : \mathbb{R}^n \to \mathbb{R}^n / \mathcal{S}^*_{g,\mathrm{im}(B)}$. Choose $D \in \mathbb{R}^{r \times n}$ such that $\ker(D) = \mathcal{S}^*_{g,\mathrm{im}(B)} \bigcap \ker(C)$ and a friend

$J \in \boldsymbol{J}(\mathcal{S}^*_{g,\mathrm{im}(B)})$ such that $\sigma\left((A+JC)/\mathcal{S}^*_{g,\mathrm{im}(B)}\right) \subset \mathbb{C}_g$. Define

$$
\begin{aligned}
E &\triangleq P(A+JC)P^+, \qquad & F &\triangleq -PJ,\\
R &\triangleq DP^+, & S &\triangleq D(I_n - P^+P)C^+.
\end{aligned}
$$

Note that in Step (S5) above, restricting $J \in \mathbb{R}^{n\times q}$ to be contained in $\boldsymbol{J}(\mathcal{S}^*_{g,\mathrm{im}(B)})$ is a matter of solving a set of linear equations while choosing $J \in \boldsymbol{J}(\mathcal{S}^*_{g,\mathrm{im}(B)})$ to assign the spectrum of $P(A+JC)P^+$ is achieved using standard pole placement techniques.

The matrices defined in the last Step (S5) of Algorithm 4.8 above indeed yield the observer we are looking for as stated in the theorem below.

Theorem 4.9. *Let system (4.4) be described by matrices $A \in \mathbb{R}^{n\times n}$, $B \in \mathbb{R}^{n\times p}$, and $C \in \mathbb{R}^{q\times p}$. Let $D \in \mathbb{R}^{r\times n}$, $E \in \mathbb{R}^{s\times s}$, $F \in \mathbb{R}^{s\times q}$, $R \in \mathbb{R}^{r\times s}$, and $S \in \mathbb{R}^{r\times q}$ be determined by the above Algorithm 4.8.*

Then the observer (4.8) is such that $\hat{\zeta}_k(t)$ is an asymptotic estimate for $\zeta_k(t) = D\rho_k(t)$ and the observer error converges to zero exponentially fast with convergence rate χ independent of initial conditions $\rho_k(0) \in \mathbb{R}^n$, $z_k(0) \in \mathbb{R}^s$.

Proof. Consider the observer error $\varepsilon_k(t) = z_k(t) - P\rho_k(t)$. The error dynamics reads

$$
\begin{aligned}
\dot{\varepsilon}_k(t) &= P(A+JC)P^+z_k(t) - PJC\rho_k(t) - P(A\rho_k(t) + B\gamma_k(t))\\
&= P(A+JC)(P^+z_k(t) - \rho_k(t)).
\end{aligned}
$$

Note that $P : \mathbb{R}^n \to \mathbb{R}^n/\mathcal{S}^*_{g,\mathrm{im}(B)}$ is the canonical projection and $P(A+JC)P^+$ is a matrix representation of the map $(A+JC)/\mathcal{S}^*_{g,\mathrm{im}(B)}$, i.e., $P(A+JC) = P(A+JC)P^+P$ and we obtain $\dot{\varepsilon}_k(t) = P(A+JC)P^+\varepsilon_k(t)$.

By definition of P and D, we have $\ker(D) = \ker(P)\bigcap\ker(C)$, i.e., $D(I_n - P^+P)C^+C = D(I_n - P^+P)$. Thus, we obtain

$$
\begin{aligned}
\hat{\zeta}_k(t) - \zeta_k(t) &= Rz_k(t) + SC\rho_k(t) - D\rho_k(t)\\
&= Rz_k(t) + D(I_n - P^+P)C^+C\rho_k(t) - D\rho_k(t)\\
&= Rz_k(t) - DP^+P\rho_k(t)\\
&= R\varepsilon_k(t).
\end{aligned}
$$

To summarize, $\sigma\left((A+JC)/\mathcal{S}^*_{g,\mathrm{im}(B)}\right) \subset \mathbb{C}_g$ is equivalent to $P(A+JC)P^+ + \chi I_s$ being Hurwitz, which implies $\|\varepsilon_k(t)\| \le Me^{-\chi t}\|\varepsilon_k(0)\|$ for some $M \in \mathbb{R}_+$. This in turn implies $\|\hat{\zeta}_k(t) - \zeta_k(t)\| = \|R\varepsilon_k(t)\| \le M\|R\|e^{-\chi t}\|\varepsilon_k(0)\|$. □

By Theorem 4.9, we thus know how to construct a maximal unknown-input observer (4.8) for the system (4.4). As a next step, we need to use the estimate obtained from this observer to achieve consensus. This will be addressed in the next section.

4.3.2 Diffusive Couplings with Reduced State Information

Since the observer (4.8) designed in the previous section is independent of the system input, the separation principle holds despite unknown parts of the input. Assume, the observer (4.8) is designed as described in Algorithm 4.8 in Section 4.3.1. With $\varepsilon_k(t) = z_k(t) - P\rho_k(t)$, systems (4.1) together with observers (4.8) and inputs $u_k(t) = \tilde{K}\hat{\zeta}_k(t) = \tilde{K}D\rho_k(t) + \tilde{K}R\varepsilon_k(t)$ for some $\tilde{K} \in \mathbb{R}^{p\times r}$ can be written as

$$\begin{pmatrix} \dot{x}_k(t) \\ \dot{\varepsilon}_k(t) \end{pmatrix} = \begin{pmatrix} A & B\tilde{K}R \\ 0 & E \end{pmatrix} \begin{pmatrix} x_k(t) \\ \varepsilon_k(t) \end{pmatrix} + \begin{pmatrix} B \\ 0 \end{pmatrix} \tilde{K}D\rho_k(t)$$

for $k \in \mathbb{N}_N$. These systems are equivalent modulo uncontrollable, asymptotically stable parts (namely the observer error dynamics) to systems (4.1) with diffusive couplings $u_k(t) = \tilde{K}D\rho_k(t)$. Therefore, the problem to solve is the State Consensus Design Problem 3.4 from Chapter 3 with $K \in \mathbb{R}^{p\times n}$ constraint to be of the form $K = \tilde{K}D$, $\tilde{K} \in \mathbb{R}^{p\times r}$. That is, by Theorem 3.7, we need to find $\tilde{K} \in \mathbb{R}^{p\times r}$ such that the matrices $A + \lambda_k(L)B\tilde{K}D$, $k \in \mathbb{N}_N \setminus \{1\}$ are Hurwitz, i.e., a static output feedback problem for the virtual output $\zeta_k(t) = D\rho_k(t)$.

Similar problems are dealt with in Tuna (2008b), where the focus is on neutrally stable systems. In what follows, we adapt the method proposed in Cao et al. (1998) and modify the LMI-based design proposed in Theorem 3.17 to obtain an iterated LMI condition for the coupling gain $\tilde{K}$.

For that purpose, we first need some preliminary definitions and results. For fixed system matrices $A \in \mathbb{R}^{n\times n}$, $B \in \mathbb{R}^{n\times p}$, and $D \in \mathbb{R}^{r\times n}$, define the complex matrices

$$\Gamma_1 \triangleq (A+\chi I)^T P + P(A+\chi I) + XBB^TX - XBB^TP - PBB^TX \in \mathbb{C}^{n\times n} \quad (4.9a)$$

$$\Gamma_2(\lambda) \triangleq PB + (\lambda\tilde{K}D)^H \in \mathbb{C}^{n\times n} \quad (4.9b)$$

$$\Gamma(\lambda) \triangleq \begin{pmatrix} \Gamma_1 & \Gamma_2(\lambda) \\ \Gamma_2^H(\lambda) & -I \end{pmatrix} \in \mathbb{C}^{2n\times 2n} \quad (4.9c)$$

in dependence of the matrices $P = P^H \in \mathbb{C}^{n\times n}$, $X = X^H \in \mathbb{C}^{n\times n}$, and $\tilde{K} \in \mathbb{R}^{p\times r}$ and the scalars $\chi \in \mathbb{R}$ and $\lambda \in \mathbb{C}$. In what follows, the above matrices are related to the static output feedback problem we are trying to solve.

Lemma 4.10. *Let $A \in \mathbb{R}^{n\times n}$, $B \in \mathbb{R}^{n\times p}$, $D \in \mathbb{R}^{r\times n}$, $\lambda \in \mathbb{C}$, and $\chi \in \mathbb{R}$.*

Given a feedback gain $\tilde{K} \in \mathbb{R}^{p\times r}$, we have $\sigma(A+\chi I+\lambda B\tilde{K}D) \subset \mathbb{C}_-$ if and only if there exist matrices $P = P^H \in \mathbb{C}^{n\times n}$ with $P \succ 0$ and $X = X^H \in \mathbb{C}^{n\times n}$ with $X \succ 0$ such that $\Gamma(\lambda) \prec 0$, with $\Gamma(\lambda)$ defined by (4.9).

Proof. The condition $\sigma(A+\chi I+\lambda B\tilde{K}D) \subset \mathbb{C}_-$ is equivalent to existence of some matrix $P = P^H \in \mathbb{C}^{n\times n}$ with $P \succ 0$ such that $\tilde{\Gamma}(\lambda) \prec 0$ with

$$\tilde{\Gamma}(\lambda) \triangleq (A+\chi I+\lambda B\tilde{K}D)^H P + P(A+\chi I+\lambda B\tilde{K}D).$$

We will establish equivalence of the above condition and the necessary and sufficient conditions given in the lemma.

($\Rightarrow$) Using the Schur complement (see Boyd et al., 1994), $\Gamma(\lambda) \prec 0$ is equivalent to $\Gamma_1 + \Gamma_2^H(\lambda)\Gamma_2(\lambda) \prec 0$. Note that $\Gamma_1 + \Gamma_2^H(\lambda)\Gamma_2(\lambda) = \tilde{\Gamma}(\lambda) + |\lambda|^2(\tilde{K}D)^T\tilde{K}D + (X - P)BB^T(X-P) \succeq \tilde{\Gamma}(\lambda)$ for any $X = X^H \in \mathbb{C}^{n\times n}$. Thus $\Gamma(\lambda) \prec 0$ implies $\tilde{\Gamma}(\lambda) \prec 0$.

($\Leftarrow$) Assume $P = P^H \in \mathbb{C}^{n\times n}$ is such that $P \succ 0$ and $\tilde{\Gamma}(\lambda) \prec 0$. Then we may assume without loss of generality that $\tilde{\Gamma}(\lambda) \prec -|\lambda|^2(\tilde{K}D)^T\tilde{K}D$. With $X = P$, we obtain $\Gamma_1 + \Gamma_2^H(\lambda)\Gamma_2(\lambda) = \tilde{\Gamma}(\lambda) + |\lambda|^2(\tilde{K}D)^T\tilde{K}D \prec 0$. Thus $\tilde{\Gamma}(\lambda) \prec 0$ implies existence of some matrix $X = X^H \in \mathbb{C}^{n\times n}$ with $X \succ 0$ such that $\Gamma(\lambda) \prec 0$. □

The matrix $\Gamma(\lambda)$ defined by (4.9) and involved in the conditions of Lemma 4.10 above is quadratic in the auxiliary matrix variable $X = X^H \in \mathbb{C}^{n\times n}$ and bilinear in X and $P = P^H \in \mathbb{C}^n$. Yet, it possesses some convenient properties. Namely, for fixed X, it is linear in the matrix variables $P = P^H \in \mathbb{C}^{n\times n}$ and $\tilde{K} \in \mathbb{R}^{p\times r}$. Furthermore, the parameter $\lambda \in \mathbb{C}$ enters Γ affinely and $\Gamma(\lambda) = \Gamma^T(\overline{\lambda})$. With these observations, we are ready to relate Lemma 4.10 to the consensus problem as follows:

Theorem 4.11. *Let $\mathcal{G} = \{\mathcal{V}, \mathcal{E}, W\}$ be a constant communication graph with $|\mathcal{V}| = N > 1$ and Laplacian matrix L satisfying Assumption 2.2. Let $A \in \mathbb{R}^{n\times n}$, $B \in \mathbb{R}^{n\times p}$, $C \in \mathbb{R}^{q\times n}$, and $\chi \in \mathbb{R}_+$. Let $\mathcal{S}^*_{g,\mathrm{im}(B)}$ be the complementary detectability subspace of (4.4) which is minimal subject to containing $\mathrm{im}(B)$ with the good part $\mathbb{C}_g$ of the complex plane defined as $\mathbb{C}_g \triangleq \{z \in \mathbb{C} | \mathfrak{Re}(z) < -\chi\}$. Let $r = n - \dim\left(\mathcal{S}^*_{g,\mathrm{im}(B)} \bigcap \ker(C)\right)$ and $D \in \mathbb{R}^{r\times n}$ such that $\ker(D) = \mathcal{S}^*_{g,\mathrm{im}(B)} \bigcap \ker(C)$. Let $s \in \mathbb{N}$ and $\mu_j \in \mathbb{C}$, $j \in \mathbb{N}_s$ such that $\lambda_k \in \mathrm{conv}(\{\mu_1, \dots, \mu_s, \overline{\mu}_1, \dots, \overline{\mu}_s\}) \subset \mathbb{C}$ for all $k \in \mathbb{N}_N \setminus \{1\}$.*

If there exist matrices $P = P^H \in \mathbb{C}^{n\times n}$, $X = X^H \in \mathbb{C}^{n\times n}$ and $\tilde{K} \in \mathbb{R}^{p\times r}$ such that $P \succ 0$, $X \succ 0$ and $\Gamma(\mu_j) \prec 0$, $j \in \mathbb{N}_s$, then the dynamic couplings (4.3) with $G = \tilde{K}R$ and $H = \tilde{K}S$ and matrices E, F, R, S determined in Algorithm 4.8 in Section 4.3.1 ensure that output consensus is reached exponentially fast with convergence bound proportional to $e^{-\chi t}$.

Proof. We know that the separation principle holds. Furthermore, by Theorem 4.9, we know that the observer errors $\varepsilon_k(t)$ satisfy $\|\varepsilon_k(t)\| \leq Me^{-\chi t}\|\varepsilon_k(0)\|$. It remains to show that $\sigma(A + \lambda_k B\tilde{K}D) \subset \mathbb{C}_g$ for all $k \in \mathbb{N}_N \setminus \{1\}$. Note that $\Gamma(\mu_1) \prec 0$ and $\Gamma(\mu_2) \prec 0$ by convexity of $\Gamma(\cdot)$ implies that $\Gamma(\theta\mu_1 + (1-\theta)\mu_2) = \theta\Gamma(\mu_1) + (1-\theta)\Gamma(\mu_2) \prec 0$ for any $\theta \in [0,1]$. Furthermore $\Gamma(\mu) \prec 0$ is equivalent to $\Gamma(\overline{\mu}) = \Gamma^T(\mu) \prec 0$. Thus since $\lambda_k(L) \in \mathrm{conv}(\{\mu_1, \dots, \mu_s, \overline{\mu}_1, \dots, \overline{\mu}_s\})$ for $k \in \mathbb{N}_N \setminus \{1\}$, $\Gamma(\mu_j) \prec 0$, $j \in \mathbb{N}_s$ implies $\Gamma(\lambda_k(L)) \prec 0$, $k \in \mathbb{N}_N \setminus \{1\}$. The theorem then follows by Lemma 4.10. □

Theorem 4.11 above can be used to determine the coupling gain $\tilde{K} \in \mathbb{R}^{p\times r}$ by solving a set of matrix inequalities depending on complex numbers μ_j, $j \in \mathbb{N}_s$ whose convex hull contains the non-zero eigenvalues of the Laplacian matrix L[1]. As mentioned before, those matrix inequalities involve quadratic and bilinear terms in the unknowns, i.e., they are not solvable using standard LMI solvers. However, it is possible to solve the matrix inequalities by iteratively solving LMIs as described in the following algorithm adapted from Cao et al. (1998):

Algorithm 4.12.

(S1) Set $\nu = 1$ and $X_1 \in \mathbb{R}^{n\times n}$ the unique positive definite solution to the algebraic Riccati equation $A^TX_1 + X_1A - X_1BB^TX_1 + Q = 0$ for some $Q = Q^T \in \mathbb{R}^{n\times n}$ with $Q \succ 0$.

[1] A method to determine the values $\mu_j \in \mathbb{C}$, $j \in \mathbb{N}_s$ was given in Corollary 3.19.

(S2) Solve the optimization problem

$$\begin{aligned} \min_{P_\nu, \tilde{K}_\nu, \chi_\nu} \ & \chi_\nu \\ \text{s.t.} \quad & P_\nu = P_\nu^H \succ 0 \\ & \Gamma(\mu_j) \prec 0, \quad j \in \mathbb{N}_s, \end{aligned}$$

with $P = P_\nu$, $X = X_\nu$, $\tilde{K} = \tilde{K}_\nu$, and $\chi = \chi_\nu$ substituted into $\Gamma(\cdot)$, for $P_\nu \in \mathbb{C}^{n\times n}$, $\tilde{K}_\nu \in \mathbb{R}^{p\times r}$, and $\chi_\nu \in \mathbb{R}$ performing a bisection on χ_ν. Denote the optimal value as χ_ν^*.

(S3) If $\chi_\nu^* \leq \chi$, the algorithm stops and $P = P_\nu$, $X = X_\nu$ and $\tilde{K} = \tilde{K}_\nu$ solve the original matrix inequalities from Theorem 4.11.

(S4) Solve the optimization problem

$$\begin{aligned} \min_{P'_\nu, \tilde{K}'_\nu} \ & \operatorname{trace}(P'_\nu) \\ \text{s.t.} \quad & P'_\nu = (P'_\nu)^H \succ 0 \\ & \Gamma(\mu_j) \prec 0, \quad j \in \mathbb{N}_s, \end{aligned}$$

with $P = P'_\nu$, $X = X_\nu$, $\tilde{K} = \tilde{K}'_\nu$, and $\chi = \chi_\nu^*$ substituted into $\Gamma(\cdot)$, for $P'_\nu \in \mathbb{C}^{n\times n}$ and $\tilde{K}'_\nu \in \mathbb{R}^{p\times r}$. Denote the optimum as P_ν^*.

(S5) If $\|X_\nu - P_\nu^*\| < \delta$ – a prescribed tolerance – the iteration stops without solution; else set $X_{\nu+1} = P_\nu^*$ and $\nu = \nu + 1$ and go to Step (S2).

For a proof of convergence of Algorithm 4.12 and a discussion of some of its features, the reader is referred to Cao et al. (1998). It should be noted that the choice of the iterated LMI condition proposed in Cao et al. (1998) is to some extent arbitrary. It could be replaced by other approaches from static output stabilization like Ghaoui et al. (1997), Leibfritz (2001), Mangasarian and Pang (1995).

4.3.3 Example

As an illustrative example, we choose a scenario similar to Section 3.2.2, i.e., we consider again a formation problem for a group of $N = 9$ vehicles moving on a plane. We choose the same communication graph and target formation as in Section 3.2.2 (see Figure 3.3). We refrain from giving further simulation figures in the present example, since the results would look very similar to those given in Figure 3.4 before. The main objective of this example is to illustrate the design of the unknown-input observer (4.8) and the solution of the matrix inequalities involved in Theorem 4.11 and Algorithm 4.12.

We consider slightly modified dynamic models for the individual systems as compared to the model (3.16) used in the example in Section 3.2.2. Namely, we consider vehicles actuated by a second order dynamical system and we assume that we can measure the relative position and the relative first actuator state of the individual vehicles. The resulting individual systems model is given by (4.1) with

$$A = \begin{pmatrix} 0 & 1 & 0 & 0 \\ 0 & -\alpha_1 & 1 & 0 \\ 0 & 0 & -\alpha_2 & 1 \\ 0 & 0 & 0 & -\alpha_3 \end{pmatrix}, \quad B = \begin{pmatrix} 0 \\ 0 \\ 0 \\ 1 \end{pmatrix},$$
$$C = \begin{pmatrix} 1 & 0 & 0 & 0 \\ 0 & 0 & 1 & 0 \end{pmatrix}.$$

As a first step, we need to determine the maximal unknown-input observable. To that end, we determine $\mathcal{S}^*_{\mathrm{im}(B)}$, $\mathcal{M}^*_{\mathrm{im}(B)}$, and $\mathcal{S}^*_{g,\mathrm{im}(B)}$ using the algorithms given in Appendix C.2.2 as described in Steps (S1) – (S5) of Algorithm 4.8. The sequence in Algorithm C.5 reads

$$\begin{aligned} \mathcal{S}^0_{\mathrm{im}(B)} &= \{0\}, \\ \mathcal{S}^1_{\mathrm{im}(B)} &= \mathrm{im}(B), \\ \mathcal{S}^2_{\mathrm{im}(B)} &= \mathrm{im}\left(\begin{pmatrix} 0_{2\times 2} \\ I_2 \end{pmatrix}\right), \\ \mathcal{S}^3_{\mathrm{im}(B)} &= \mathcal{S}^2_{\mathrm{im}(B)} = \mathcal{S}^\infty_{\mathrm{im}(B)}, \end{aligned}$$

i.e., $\mathcal{S}^*_{\mathrm{im}(B)} = \mathcal{S}^2_{\mathrm{im}(B)}$. The sequence in Algorithm C.6 reads

$$\begin{aligned} \mathcal{M}^0_{\mathrm{im}(B)} &= \mathbb{R}^n, \\ \mathcal{M}^1_{\mathrm{im}(B)} &= \ker\left(\begin{pmatrix} 1 & 0 & 0 & 0 \end{pmatrix}\right), \\ \mathcal{M}^2_{\mathrm{im}(B)} &= \ker\left(\begin{pmatrix} I_2 & 0_{2\times 2} \end{pmatrix}\right) = \mathcal{S}^*_{\mathrm{im}(B)}, \end{aligned}$$

i.e., $\mathcal{M}^*_{\mathrm{im}(B)} = \mathcal{M}^2_{\mathrm{im}(B)} = \mathcal{S}^*_{\mathrm{im}(B)}$. Due to the inclusion relation $\mathcal{S}^*_{\mathrm{im}(B)} \subset \mathcal{S}^*_{g,\mathrm{im}(B)} \subset \mathcal{M}^*_{\mathrm{im}(B)}$, we thus know that $\mathcal{S}^*_{g,\mathrm{im}(B)} = \mathcal{M}^*_{\mathrm{im}(B)}$.

A matrix $D \in \mathbb{R}^{r\times 4}$ such that $\ker(D) = \mathcal{S}_{g,\mathrm{im}(B)} \bigcap \ker(C)$ is given as

$$D = \begin{pmatrix} I_3 & 0 \end{pmatrix}$$

with $r = 3$, i.e., we can construct an observer for the first three components of the relative states of the individual systems.

To construct the observer, we first determine the canonical projection $P : \mathbb{R}^4 \to \mathbb{R}^4/\mathcal{S}^*_{g,\mathrm{im}(B)}$ in matrix form as $P = (I_2, 0_{2\times 2})$. The set $\boldsymbol{J}(\mathcal{S}^*_{g,\mathrm{im}(B)})$ of friends of $\mathcal{S}^*_{g,\mathrm{im}(B)}$ can then be characterized as the set of matrices $J \in \mathbb{R}^{4\times 2}$ that solve $P(A+JC)(I_4 - P^+P) = 0$, i.e., matrices of the form

$$J = \begin{pmatrix} j_{1,1} & 0 \\ j_{2,1} & -1 \\ j_{3,1} & j_{3,2} \\ j_{4,1} & j_{4,2} \end{pmatrix} \in \mathbb{R}^{4\times 2}.$$

Finally, the observer matrices read

$$E = \begin{pmatrix} j_{1,1} & 1 \\ j_{2,1} & -\alpha_1 \end{pmatrix}, \quad F = \begin{pmatrix} j_{1,1} & 0 \\ j_{2,1} & -1 \end{pmatrix},$$
$$R = \begin{pmatrix} 1 & 0 \\ 0 & 1 \\ 0 & 0 \end{pmatrix}, \quad S = \begin{pmatrix} 0 & 0 \\ 0 & 0 \\ 0 & 1 \end{pmatrix},$$

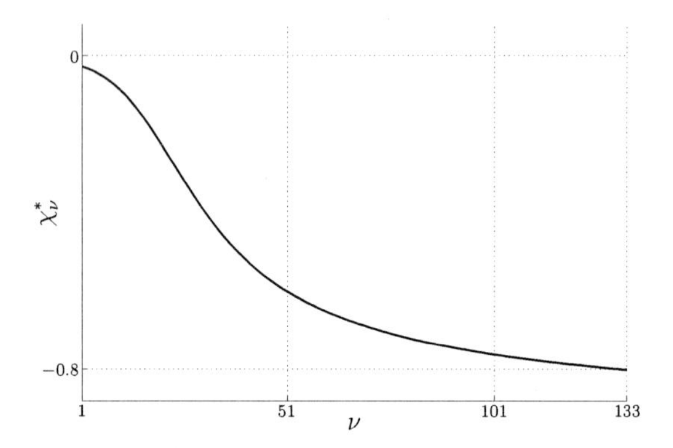

Figure 4.1: Convergence of χ^*_ν in Algorithm 4.12.

where $j_{1,1}, j_{2,1} \in \mathbb{R}$ are degrees of freedom that can be chosen to assign the spectrum of E, e.g., using standard pole placement methods. For instance, equating coefficients yields $j_{1,1} = \alpha_1 + \lambda_1(E) + \lambda_2(E)$ and $j_{2,1} = -(\alpha_1 + \lambda_1(E))(\alpha_1 + \lambda_2(E))$. Note that $j_{3,k}$ and $j_{4,k}$, $k = 1, 2$ do not influence the observer dynamics at all.

The static output feedback problem is solved using Algorithm 4.12. With $\chi = -0.8$ and $Q = \operatorname{diag}((100, 10, 2, 1))$, Algorithm 4.12 terminates after 133 iterations. The evolution of χ^*_ν in Algorithm 4.12 is depicted in Figure 4.1. Note that smaller values of χ can be achieved, but Algorithm 4.12 progresses much slower for smaller values of χ. The resulting coupling gain is $\tilde{K} \approx -(166.0, 164.2, 47.54)$.

4.4 Summary

In this chapter, we considered the problem of consensus subject to the constraint of relative output sensing, i.e., a restriction on the type of couplings between the individual systems. As a consequence of this restriction, the consensus problem has to be solved using only limited information. Namely, the only information available is that of relative output measurements. This situation corresponds to more general system descriptions, i.e., higher system complexity, in the framework of relative sensing.

We showed that relative output sensing imposes severe restrictions in case of time-varying communication topologies. In case of fixed communication topologies, i.e., constant communication graphs, the problem can be solved with the help of unknown-input observers.

We proposed solutions (see Wieland and Allgöwer, 2010) based on full order unknown input observers which come on the one hand with the advantage that the static couplings from Chapter 3 can be applied without modification by virtue of the separation principle. But those full-order unknown-input observers on the other hand come with the disadvantage of quite severe restrictions on the individual system dynamics.

Those restrictions have been partially relaxed with the help of reduced order unknown-input observers. Since, in that case, we only estimate parts of the relative states, the methods from Chapter 3 need to be adapted to take into account the constraint on the coupling gains to only depend on the parts of the states that are estimated be the observer. This yields to a static output feedback problem. An adaptation of the LMI based design methods in Chapter 3 has been proposed yielding an iterated LMI condition.

In both cases, the observer based design yields dynamic couplings, that allow for consensus among larger classes of systems, i.e., higher system complexity, than what would be possible with static couplings proposed in Chapter 3.

Chapter 5

The Internal Model Principle for Consensus and Synchronization

In the previous two chapters, as well as in the huge majority of the references given therein, the state and output consensus problem has been addressed under the assumption that all individuals in the group admit identical dynamical models. However, in a group of physical systems, there will be hardly two individuals that are exactly identical. Systems may be structurally different, e.g., due to different types of actuators, like for instance groups of mobile robots, some of which are equipped with combustion engines of different power while some others use electric motors. But even if the individuals are structurally the same, they still may exhibit non-identical behavior, e.g., due to different parameter values. There is a huge variety of possible sources of such heterogeneity in parameters. Examples include different friction or damping coefficients, different or changing masses or change of material properties due to abrasion, to mention only a few.

Despite the fact, that individual systems in a group are generally non-identical to some extent, we still observe synchronization and consensus in many cases. Therefore, we ask what properties need to be shared by the dynamic models of the individual systems in a group in order to be able to synchronize the individuals in a meaningful way. We already gave one answer to a specific instance of this problem in Section 3.4, where we considered a particular system in the group to be a leader and derived constraints on the leader dynamics that need to be satisfied in order to achieve state consensus. In fact, we showed that the leader dynamics needs to have some specific parts in common with the dynamics of the remaining systems. In what follows, we will not assume a specific communication topology, like, e.g., a specific leader, but derive results, which constrain the dynamic models of all individuals in the group.

Consensus in heterogeneous groups has been addressed before, e.g., in Qu et al. (2008). The approach chosen there uses agent level controls that transform all individuals into a canonical form which is the same for all individuals. However, such an approach is not always possible due to required system properties or communication constraints in the network. The approach presented here is based on Wieland and Allgöwer (2009a), Wieland et al. (2010c) for linear system models and Wieland and Allgöwer (2009b) for nonlinear system models.

In both cases, we will derive an *internal model principle for synchronization*. We will show that output synchronization among non-identical systems using diffusive-like couplings is possible only if all individual systems, together with their local coupling dynamics, contain an internal model of some common virtual exosystem. Moreover, if such a common virtual exosystem exists, it generates all possible synchronous outputs for the network in the linear case. In the nonlinear case additional assumptions are needed for this property to be satisfied.

The internal model requirement will be expressed in terms of linear matrix equations, known as Francis equations, in the linear case and nonlinear partial differential equations, known as FBI equations, in the nonlinear case. These equations relate the individual system dynamics to the virtual exosystem dynamics. Both types of conditions are known from the theory of output regulation of linear and nonlinear systems. It will in fact turn out that the internal model principle for synchronization is very similar to the classical internal model principle for control theory (Francis and Wonham, 1976). Yet, there are some important differences. While the classical internal model principle deals with a special case of the output regulation problem (see Byrnes et al., 1997, Knobloch et al., 1993), namely the problem of one system which shall track one specific exosystem, the situation in synchronization problems is more complex. There exists no exosystem to be tracked by individual agents. Instead, all individual systems need to mutually track each other and thereby all systems mutually influence each other.

The internal model principle for synchronization in groups of linear systems will be presented in Section 5.1. The case of networks of nonlinear systems is dealt with in Section 5.2.

5.1 The Linear Case

5.1.1 Problem Setup

The network we are considering here is similar to the one considered in Section 4.1, except that the individuals are modeled with non-identical dynamics, i.e., we consider a group of N LTI systems

$$\dot{x}_k(t) = A_k x_k(t) + B_k u_k(t) \tag{5.1a}$$
$$y_k(t) = C_k x_k(t) \tag{5.1b}$$

where $x_k(t) \in \mathbb{R}^{n_k}$ is the state vector, $u_k(t) \in \mathbb{R}^{p_k}$ is the input vector and $y_k(t) \in \mathbb{R}^q$ is the output vector for $k \in \mathbb{N}_N$. We do not require that $n_i = n_j$ or $p_i = p_j$ for $i, j \in \mathbb{N}_N, i \neq j$, i.e., the dynamics of the individual systems, described by the matrices $A_k \in \mathbb{R}^{n_k \times n_k}$, $B_k \in \mathbb{R}^{n_k \times p_k}$, and $C_k \in \mathbb{R}^{q \times n_k}$ may be different including state and input dimensions. The outputs $y_k(t) \in \mathbb{R}^q$, $k \in \mathbb{N}_N$ have the same dimension q for all individual systems. In fact the outputs all have the same physical meaning, since these are the quantities we want to compare and synchronize. That is, we are again interested in output consensus in the sense of Definition 4.1, but this time we replace the network of identical systems (4.1) by the heterogeneous network (5.1).

The couplings between the individual systems are taken more general than in the previous section as

$$\dot{z}_k(t) = E_k z_k(t) + F_k \delta_k(t) + M_k y_k(t) \tag{5.2a}$$
$$u_k(t) = G_k z_k(t) + H_k \delta_k(t) + O_k y_k(t) \tag{5.2b}$$
$$\zeta_k(t) = P_k z_k(t) + Q y_k(t) \tag{5.2c}$$
$$\delta_k(t) = \sum_{j=1}^{N} w_{k,j}(t)(\zeta_k(t) - \zeta_j(t)) \tag{5.2d}$$

with coupling states $z_k(t) \in \mathbb{R}^{m_k}$ and virtual outputs $\zeta_k(t) \in \mathbb{R}^r$ for $k \in \mathbb{N}_N$. Couplings (5.2) represent a general LTI dynamic controller driven by the system outputs $y_k(t)$ and the relative controller output signal $\delta_k(t)$. The outputs of the controller are the system input $u_k(t)$ and the virtual output $\zeta_k(t)$ obtained as a linear function of the system outputs $y_k(t)$ and the controller states $z_k(t)$. The virtual output is introduced to allow for exchange of relative system *and* controller states over the network. Since the system outputs $y_k(t)$, $k \in \mathbb{N}_N$ all have the same physical meaning, the matrix Q in (5.2c) is the same for all $k \in \mathbb{N}_N$. Similarly to previous chapters, the values $w_{k,j}(t) \in \overline{\mathbb{R}}_+$, $j, k \in \mathbb{N}_N$ are the elements of the adjacency matrix $W(t) = [w_{k,j}(t)] \in \mathbb{R}^{N \times N}$ of some communication graph $\mathcal{G}(t) = \{\mathcal{V}, \mathcal{E}(t), W(t)\}$. A block diagram of an individual system (5.1) with its local coupling dynamics (5.2a) – (5.2c) and network interconnections (5.2d) is depicted in Figure 5.1(a). Different to the dynamic couplings (4.3) considered before, we do not impose asymptotically stable coupling dynamics at this point. The reason for that will become apparent later on.

It will be convenient in what follows to write the closed loops of systems (5.1) together with their couplings (5.2) as a single system

$$\dot{x}_k^*(t) = A_k^* x_k^*(t) + B_k^* \delta_k(t) \tag{5.3a}$$
$$y_k(t) = C_k^* x_k^*(t) \tag{5.3b}$$
$$\zeta_k(t) = P_k^* x_k^*(t) \tag{5.3c}$$

for $k \in \mathbb{N}_N$ with extended state $x_k^*(t) = (x_k(t)^T, z_k(t)^T)^T \in \mathbb{R}^{n_k+m_k}$ and matrices

$$A_k^* \triangleq \begin{pmatrix} A_k + B_k O_k C_k & B_k G_k \\ M_k C_k & E_k \end{pmatrix}, \qquad B_k^* \triangleq \begin{pmatrix} B_k H_k \\ F_k \end{pmatrix},$$
$$C_k^* \triangleq \begin{pmatrix} C_k & 0 \end{pmatrix}, \qquad P_k^* \triangleq \begin{pmatrix} Q C_k & P_k \end{pmatrix}$$

for all $k \in \mathbb{N}_N$. As before, in order to stay in the framework of generalized diffusive couplings, we want the effect of the network interconnections to disappear once the systems are synchronized. That is, in what follows, we will not only require the system outputs $y_k(t)$, $k \in \mathbb{N}_N$ to asymptotically synchronize but also the virtual outputs $\zeta_k(t)$, $k \in \mathbb{N}_N$ to be asymptotically identical. In addition, we will impose the constraint of exponential convergence to the synchronous trajectories. This yields the following formal problem description:

Problem 5.1 (Linear Heterogeneous Output Synchronization). *Let the N individual systems (5.1) be modeled by given matrices $A_k \in \mathbb{R}^{n_k \times n_k}$, $B_k \in \mathbb{R}^{n_k \times p_k}$, and $C_k \in \mathbb{R}^{q \times n_k}$*

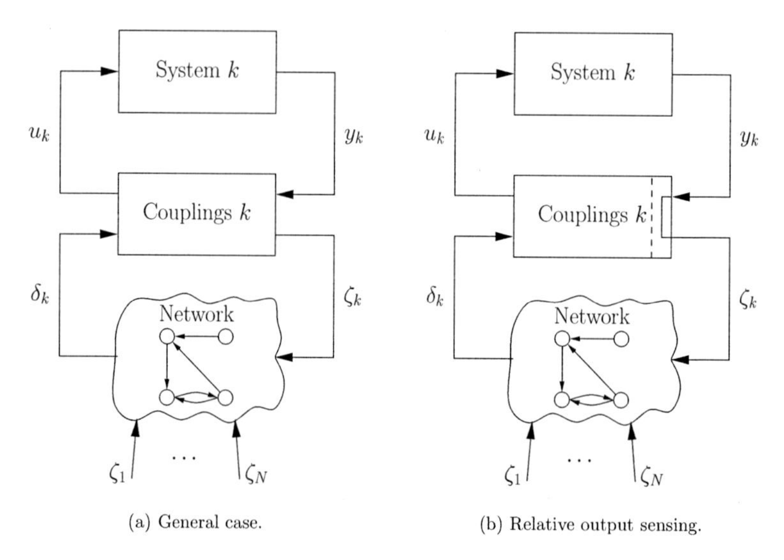

(a) General case. (b) Relative output sensing.

Figure 5.1: Block diagrams of an individual system (5.1) with local coupling dynamics (5.2a) – (5.2c) and network interconnections (5.2d) in the general case (a) and in the special case of relative output sensing (b).

for $k \in \mathbb{N}_N$. Let the communication topology in the couplings (5.2d) *be defined by some communication graph $\mathcal{G}(t) = \{\mathcal{V}, \mathcal{E}(t), W(t)\}$ with $|\mathcal{V}| = N$.*

Find, if possible, matrices $E_k \in \mathbb{R}^{m_k \times m_k}$, $F_k \in \mathbb{R}^{m_k \times r}$, $M_k \in \mathbb{R}^{m_k \times q}$, $G_k \in \mathbb{R}^{p_k \times m_k}$, $H_k \in \mathbb{R}^{p_k \times r}$, $O_k \in \mathbb{R}^{p_k \times q}$, $P_k \in \mathbb{R}^{r \times m_k}$ for $k \in \mathbb{N}_N$, and $Q \in \mathbb{R}^{r \times q}$, such that the closed loop of the N systems (5.1) *with the dynamic couplings* (5.2) *satisfies $(y_k(t) - y_j(t)) \to 0$ exponentially fast as $t \to \infty$ and $(\zeta_k(t) - \zeta_j(t)) \to 0$ exponentially fast as $t \to \infty$ for all $j, k \in \mathbb{N}_N$ and all initial conditions $x_k(0) \in \mathbb{R}^{n_k}, z_k(0) \in \mathbb{R}^{n_k}$, $k \in \mathbb{N}_N$.*

Note that, as before, asymptotic convergence in the above problem does not necessarily imply exponential convergence even though the systems are linear. The reason for that is the time-varying dynamics in (5.2d) stemming from the time-varying communication topology. We avoid these cases by making exponential convergence part of the above problem statement.

Remark 5.2. *In many instances, one is interested in particular solutions to the above problem, i.e., solutions with additional constraints imposed on the couplings* (5.2). *One case, which is often considered to be relevant, is the case of relative output sensing that was considered in Chapter 4 (see also Kim et al., 2010, Wieland and Allgöwer, 2009a). This case corresponds to a solution to the Linear Heterogeneous Output Synchronization Problem 5.1 with $M_k = 0$, $O_k = 0$, $P_k = 0$ for $k \in \mathbb{N}_N$ and $Q = I_q$. The corresponding block diagram is depicted in Figure 5.1(b). However, we are interested in* necessary *conditions for solvability of Problem 5.1. Since any necessary condition for the general case* *is also a necessary condition for special cases, we formulate the results in this section for the general couplings* (5.2).

In case of output consensus in networks of identical LTI systems, we showed in Lemma 4.3 that under some technical assumptions on the system and coupling dynamics exponential convergence to consensus implies convergence to an output trajectory of the open loop system. That is, if consensus is reached, the behavior of the group at consensus corresponds to the open loop behavior of the individual systems. We cannot hope for a similar result in the case of non-identical systems considered in this Chapter. Since there is not one open loop model of the individual systems but there are N different models (5.1), we cannot make an a priori statement about the synchronized behavior of the group. However, similarly to Definition 3.9 in Chapter 3, we want to exclude trivial consensus, i.e., the case when $y_k(t) \to 0$, $k \in \mathbb{N}_N$ as $t \to \infty$ independent of initial conditions $x_k(0) \in \mathbb{R}^{n_k}$, $z_k(0) \in \mathbb{R}^{m_k}$, $k \in \mathbb{N}_N$. In order to formulate an assumption which excludes trivial consensus, we write the global system as

$$\dot{x}^*(t) = A^* x^*(t) + B^* \delta(t) \tag{5.4a}$$

$$y(t) = C^* x^*(t) \tag{5.4b}$$

$$\zeta(t) = P^* x^*(t) \tag{5.4c}$$

$$\delta(t) = (L(t) \otimes I_r)\zeta(t) \tag{5.4d}$$

with stacked state vector $x^*(t) = (x_1^*(t)^T, \ldots, x_N^*(t)^T)^T \in \mathbb{R}^\sigma$ where $\sigma = \sum_{k=1}^N (n_k + m_k)$, stacked system output vector $y(t) = (y_1(t)^T, \ldots, y_N(t)^T)^T \in \mathbb{R}^{Nq}$, and stacked virtual output vector $\zeta(t) = (\zeta_1(t)^T, \ldots, \zeta_N(t)^T)^T \in \mathbb{R}^{Nr}$, where A^*, B^*, C^*, and P^* are block diagonal matrices obtained by stacking the systems (5.3). The couplings between the systems appear as static diffusive couplings between the virtual output $\zeta(t)$ and the input $\delta(t)$. The communication topology is encoded through the Laplacian matrix $L(t)$.

In what follows, we will call a time-varying linear map $A(t) : \mathbb{R}^\sigma \to \mathbb{R}^\sigma$ asymptotically (exponentially) stable if the corresponding dynamical system $\dot{x}(t) = A(t)x(t)$, $x(t) \in \mathbb{R}^\sigma$ is asymptotically (exponentially) stable, i.e., if the corresponding fundamental map $\Phi(t, t_0) : \mathbb{R}^\sigma \to \mathbb{R}^\sigma$ satisfies $\|\Phi(t, t_0)\| \to 0$ as $t \to \infty$ ($\|\Phi(t, t_0)\| \leq M e^{-\mu(t-t_0)}$, $t > t_0$ for some $M, \mu \in \mathbb{R}_+$).

The assumption that excludes trivial consensus reads as follows:

Assumption 5.3. *The pair* (C^*, A^*) *is detectable and the closed loop matrix* $A_{\text{cl}}^*(t) \triangleq A^* + B^*(L(t) \otimes I_r)P^*$ *is* not *asymptotically stable.*

In simple words, Assumption 5.3 states that the origin is not an asymptotically stable equilibrium for the closed loop system (5.4). If that was the case, consensus would clearly be trivial. In addition to non-decaying state trajectories $x^*(t)$ for system (5.4), we need that the state trajectories are visible at the output $y(t)$. This is guaranteed by assuming detectability of the pair (C^*, A^*). Detectability of (C^*, A^*) is actually slightly more than what we need to exclude trivial consensus. It would be enough to guarantee existence of some solutions $x^*(t)$ of (5.4) such that $y(t) = C^* x^*(t)$ does not converge to the origin. The reason for the stronger assumption is twofold: firstly, it is probably not desirable that parts of the system state $x^*(t)$ grow unbounded without being measurable at the output and secondly, detectability is a technical assumption needed in the presentation of the results below. In addition, assuming detectability of the system (5.4a), (5.4b) or equivalently the systems (5.3a), (5.3b) seems generally not to be a severe restriction.

In what follows, we will not aim for a solution to the Linear Heterogeneous Output Synchronization Problem 5.1 but instead answer the question what can be deduced about the individual systems and their dynamic coupling controllers if we know that there exists a solution to Problem 5.1. That is, we will derive *necessary conditions* for solvability of Problem 5.1.

5.1.2 The Internal Model Principle for Synchronization among Linear Systems

The objective in this section is to derive conditions on the problem data, i.e., the description of the models of the individual systems (5.1), that are necessary for solvability of the Linear Heterogeneous Output Synchronization Problem 5.1. As a first step, we derive necessary conditions that are implicit in the sense that they depend not only on the problem data but also on the solution to Problem 5.1, i.e the dynamic coupling controllers (5.2). Once this result is established, it will however be easy to obtain explicit conditions, that depend on the problem data only.

Implicit Internal Model Principle

We state the first version of the necessary conditions for solvability of the Linear Heterogeneous Output Synchronization Problem 5.1 as follows:

Theorem 5.4. *Let $\mathcal{G}(t) = \{\mathcal{V}, \mathcal{E}(t), W(t)\}$ be some communication graph with $|\mathcal{V}| = N > 1$ and Laplacian matrix $L(t)$. Let $A_k \in \mathbb{R}^{n_k \times n_k}$, $B_k \in \mathbb{R}^{n_k \times p_k}$, and $C_k \in \mathbb{R}^{q \times n_k}$ for $k \in \mathbb{N}_N$.*

If the matrices $E_k \in \mathbb{R}^{m_k \times m_k}$, $F_k \in \mathbb{R}^{m_k \times r}$, $M_k \in \mathbb{R}^{m_k \times q}$, $G_k \in \mathbb{R}^{p_k \times m_k}$, $H_k \in \mathbb{R}^{p_k \times r}$, $O_k \in \mathbb{R}^{p_k \times q}$, $P_k \in \mathbb{R}^{r \times m_k}$ for $k \in \mathbb{N}_N$, and the matrix $Q \in \mathbb{R}^{r \times q}$ solve the Linear Heterogeneous Output Synchronization Problem 5.1 while satisfying Assumption 5.3, then

(a) there exist a positive integer $\nu \in \mathbb{N}$ and matrices $\Psi_k \in \mathbb{R}^{(n_k+m_k)\times \nu}$ with $\operatorname{rank}(\Psi_k) = \nu$ for $k \in \mathbb{N}_N$, $S \in \mathbb{R}^{\nu \times \nu}$, and $R \in \mathbb{R}^{q \times \nu}$ such that $\sigma(S) \subset \overline{\mathbb{C}}_+$, the pair (R, S) is observable, and

$$A_k^* \Psi_k = \Psi_k S \tag{5.5a}$$
$$C_k^* \Psi_k = R \tag{5.5b}$$

for all $k \in \mathbb{N}_N$; and

(b) there exists a vector $\xi_0 \in \mathbb{R}^\nu$, depending on initial conditions $x_k^*(0) \in \mathbb{R}^{n_k+m_k}$, $k \in \mathbb{N}_N$ and the communication graph $\mathcal{G}(t)$, and constants $M, \mu \in \mathbb{R}_+$ such that

$$\left\| y_k(t) - Re^{St}\xi_0 \right\| \leq Me^{-\mu t} \max_{j \in \mathbb{N}_N} \|y_j(0) - R\xi_0\|, \quad t \geq 0 \tag{5.6}$$

for all $k \in \mathbb{N}_N$.

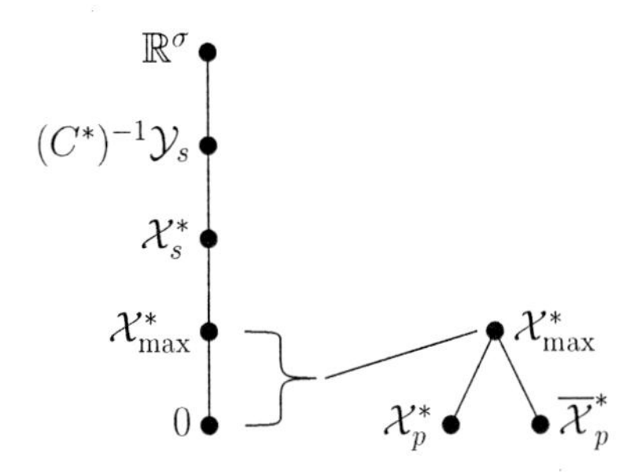

Figure 5.2: Lattice diagrams for subspace inclusions of subspaces needed for the proof of Theorem 5.4.

To prove Theorem 5.4 we need to define several subspaces. First, we define the *synchronous system output subspace*

$$\mathcal{Y}_s \triangleq \left\{ y = (y_1^T, \cdots, y_N^T)^T \in \mathbb{R}^{Nq} | y_1 = \cdots = y_N \in \mathbb{R}^q \right\} \subset \mathbb{R}^{Nq}$$

where all individual system outputs $y_k(t)$, $k \in \mathbb{N}_N$ have the same value and the *synchronous virtual output subspace*

$$\mathcal{W}_s \triangleq \left\{ \zeta = (\zeta_1^T, \cdots, \zeta_N^T)^T \in \mathbb{R}^{Nr} | \zeta_1 = \cdots = \zeta_N \in \mathbb{R}^r \right\} \subset \mathbb{R}^{Nr}$$

where all individual virtual outputs $\zeta_k(t)$, $k \in \mathbb{N}_N$ have the same value. If the Linear Heterogeneous Output Synchronization Problem 5.1 is solved, the outputs $y(t)$ and $\zeta(t)$ of (5.4) asymptotically converge to these subspaces, i.e. the state $x^*(t)$ asymptotically converges to the *synchronous subspace*

$$\mathcal{X}_s^* \triangleq \left((C^*)^{-1}\mathcal{Y}_s\right) \bigcap \left((P^*)^{-1}\mathcal{W}_s\right) \subset \mathbb{R}^\sigma.$$

The largest subspace of $\mathbb{R}^\sigma$ contained in $\mathcal{X}_s^*$, which is invariant under A^*, called the *maximal invariant synchronous subspace*, is denoted as $\mathcal{X}_{\max}^*$. This subspace is well-defined; yet, it may be the trivial subspace without further assumptions on the matrix A^* and the subspace $\mathcal{X}_s^*$. The closed loop dynamics of the global system (5.4) is governed by the time-varying matrix $A_{\text{cl}}^*(t) \triangleq A^* + B^*(L(t) \otimes I_r)P^*$. Since $A_{\text{cl}}^*(t)x^* = A^*x^*$ for all $x^* \in \mathcal{X}_s^*$ and all $t \in \mathbb{R}$ by definition of $\mathcal{X}_s^*$, $\mathcal{X}_{\max}^*$ is also invariant under $A_{\text{cl}}^*(t)$ and is the largest subspace of $\mathcal{X}_s^*$ with that property. Finally, let $\mathcal{X}_p^*, \overline{\mathcal{X}}_p^* \subset \mathcal{X}_{\max}^*$ be subspaces such that $\mathcal{X}_{\max}^* = \mathcal{X}_p^* \oplus \overline{\mathcal{X}}_p^*$, $A^*\mathcal{X}_p^* \subset \mathcal{X}_p^*$, $A^*\overline{\mathcal{X}}_p^* \subset \overline{\mathcal{X}}_p^*$, the spectrum of A^* restricted to $\mathcal{X}_p^*$ is contained in the closed right-half complex plane, i.e., $\sigma(A^*|\mathcal{X}_p^*) \subset \overline{\mathbb{C}}_+$, and the spectrum of A^* restricted to $\overline{\mathcal{X}}_p^*$ is contained in the open left-half complex plane, i.e., $\sigma(A^*|\overline{\mathcal{X}}_p^*) \subset \mathbb{C}_-$. We call $\mathcal{X}_p^*$ the *persistent invariant synchronous subspace.* The inclusion relations between the subspaces thus defined are illustrated in the lattice diagrams in Figure 5.2.

Proof of Theorem 5.4. We start by showing that any solution of (5.4) converges exponentially fast to a solution of (5.4) contained in the persistent invariant synchronous subspace

$\mathcal{X}_p^* \subset \mathbb{R}^\sigma$, i.e., in particular, $A_{\text{cl}}^*(t)/\mathcal{X}_p^*$ is exponentially stable (see Appendix C for a definition of the induced map $A_{\text{cl}}^*(t)/\mathcal{X}_p^*$). Furthermore, $\mathcal{X}_p^*$ is the smallest subspace with this property.

Suppose $(y_k(t) - y_j(t)) \to 0$ and $(\zeta_k(t) - \zeta_j(t)) \to 0$ exponentially fast as $t \to \infty$ for all $j, k \in \mathbb{N}_N$. Then, $\inf_{\check{x}^* \in \mathcal{X}_s^*} \|x^*(t) - \check{x}^*\| \to 0$ exponentially fast as $t \to \infty$ for any solution $x^*(t)$ of (5.4). Since $L(t)1_N = 0$ for all $t \in \mathbb{R}$, we have $(L(t) \otimes I_r)P^*\mathcal{X}_s^* = \{0\}$ and thus $\delta(t) = (L(t) \otimes I_r)P^*x^*(t) \to 0$ exponentially fast as $t \to \infty$. Using Lemma B.1, we deduce that there exists some $\xi_c \in \mathbb{R}^\sigma$ such that $(x^*(t) - e^{A^*t}\xi_c) \to 0$ exponentially fast as $t \to \infty$, and consequently $\inf_{\check{x}^* \in \mathcal{X}_s^*} \|e^{A^*t}\xi_c - \check{x}^*\| \to 0$ exponentially fast as $t \to \infty$. Since $e^{A^*t}\xi_c$ converges to the synchronous subspace $\mathcal{X}_s^*$ exponentially fast, it converges to an invariant subspace of $\mathcal{X}_s^*$ exponentially fast, i.e., there exists $\xi_{\max} \in \mathcal{X}_{\max}^*$ such that $e^{A^*t}(\xi_c - \xi_{\max}) \to 0$ exponentially fast as $t \to \infty$. Furthermore, by definition of the persistent invariant synchronous subspace $\mathcal{X}_p^*$ and its complement $\overline{\mathcal{X}}_p^*$, there exists $\xi_0 \in \mathcal{X}_p^*$ such that $\xi_{\max} - \xi_0 \in \overline{\mathcal{X}}_p^*$ and thus $e^{A^*t}(\xi_{\max} - \xi_0) \to 0$ exponentially fast as $t \to \infty$. In summary, for any solution $x^*(t)$ to (5.4), there exists $\xi_0 \in \mathcal{X}_p^*$ such that $(x^*(t) - e^{A^*t}\xi_0) \to 0$ exponentially fast as $t \to \infty$. Since $\sigma(A^*|\mathcal{X}_p^*) \subset \overline{\mathbb{C}}_+$, $\mathcal{X}_p^*$ is the smallest subspace satisfying this property.

Next, we exploit properties of the persistent invariant synchronous subspace $\mathcal{X}_p^*$ to prove the statement (a) of the theorem.

Since $A_{\text{cl}}^*(t)/\mathcal{X}_p^*$ is exponentially stable and $A_{\text{cl}}^*(t)$ is not asymptotically stable by Assumption 5.3, we know that $\mathcal{X}_p^* \subset \mathbb{R}^\sigma$ is a non-trivial subspace. Hence, $\nu \triangleq \dim(\mathcal{X}_p^*) > 0$. Furthermore, since $\sigma(A^*|\mathcal{X}_p^*) \subset \overline{\mathbb{C}}_+$ and (C^*, A^*) is detectable by Assumption 5.3, we know that $\mathcal{M} \bigcap \mathcal{X}_p^* = \{0\}$, where $\mathcal{M} \triangleq \bigcap_{k=0}^{\sigma-1} \ker(C^*A^*)$ is the unobservable subspace of (5.4a), (5.4b). Let $S \triangleq A^*|\mathcal{X}_p^* : \mathcal{X}_p^* \to \mathcal{X}_p^*$ be the restriction of A^* to $\mathcal{X}_p^*$ and define $\mathfrak{R} \triangleq C^*|\mathcal{X}_p^* : \mathcal{X}_p^* \to \mathcal{Y}_s$ as $x \mapsto C^*x$, $x \in \mathcal{X}_p^*$. Let $\Psi : \mathcal{X}_p^* \to \mathbb{R}^\sigma$ be the injection $x \mapsto x$. Then $A^*\Psi = \Psi S$ and $C^*\Psi = \mathfrak{R}$. Since $\mathcal{M} \bigcap \mathcal{X}_p^* = \{0\}$, the pair $(\mathfrak{R}, S)$ is observable. Furthermore, the spectrum of S satisfies $\sigma(S) \subset \overline{\mathbb{C}}_+$ by definition of the persistent invariant synchronous subspace $\mathcal{X}_p^*$. Define the projections $\Xi_k : \mathbb{R}^\sigma \to \mathbb{R}^{n_k+m_k}$, $k \in \mathbb{N}_N$ as $x^* = ((x_1^*)^T, \ldots, (x_N^*)^T)^T \mapsto x_k^*$ and let $\Psi_k \triangleq \Xi_k\Psi : \mathcal{X}_p^* \to \mathbb{R}^{n_k+m_k}$, $k \in \mathbb{N}_N$. Note that $\Xi_k A^* = A_k^*\Xi_k$ and thus $A_k^*\Psi_k = \Psi_k S$, i.e., Ψ_k solves (5.5a). Furthermore, define the projections $\Upsilon_k : \mathbb{R}^{Nq} \to \mathbb{R}^q$, $k \in \mathbb{N}_N$ as $y = (y_1^T, \ldots, y_N^T)^T \mapsto y_k$. Since $\operatorname{im}(\mathfrak{R}) \subset \mathcal{Y}_s$, we have $\Upsilon_1\mathfrak{R} = \cdots = \Upsilon_N\mathfrak{R}$ and can thus define $R \triangleq \Upsilon_k\mathfrak{R}$. Note that $\Upsilon_k C^* = C_k^*\Xi_k$ and thus $C_k^*\Psi_k = R$, i.e., Ψ_k also solves (5.5b). Since $\Upsilon_k\mathfrak{R} = (\Upsilon_k|\mathcal{Y}_s)\mathfrak{R}$ and the maps $\Upsilon_k|\mathcal{Y}_s : \mathcal{Y}_s \to \mathbb{R}^q$, $k \in \mathbb{N}_N$ are isomorphisms, observability of the pair $(\mathfrak{R}, S)$ is equivalent to observability of the pair (R, S). As a consequence of (5.5) and observability of the pair (R, S), the maps Ψ_k have full rank, for $RS^j = C_k^*(A_k^*)^j\Psi_k$ for all $k \in \mathbb{N}_N$ and all $j \in \mathbb{N} \bigcup \{0\}$, i.e.,

$$\bigcap_{j=0}^{\nu-1} \ker(RS^j) = \Psi_k^{-1}\left(\bigcap_{j=0}^{\nu-1} \ker(C_k^*(A_k^*)^j)\right) = \{0\}, \quad k \in \mathbb{N}_N.$$

This completes the proof of statement (a).

It remains to prove statement (b). Recall that the map $A_{\text{cl}}^*/\mathcal{X}_p^*$ is exponentially stable, i.e., $\inf_{\check{x}^* \in \mathcal{X}_p^*} \|x^*(t) - \check{x}^*\| \to 0$ exponentially fast as $t \to \infty$ for any solution $x^*(t)$ of

(5.4). The solutions $\xi(t)$ contained in $\mathcal{X}_p^*$ satisfy $\dot{\xi}(t) = S\xi(t)$, $y_k(t) = R\xi(t)$ for $k \in \mathbb{N}_N$. Inequality (5.6) then follows from Lemma B.1. This completes the proof. □

Before refining the result from Theorem 5.4 above, some comments and interpretations are in order. We showed that the persistent invariant synchronous subspace $\mathcal{X}_p^*$ is non-trivial with $\dim(\mathcal{X}_p^*) = \nu > 0$. Furthermore, $\mathcal{X}_p^*$ is exponentially attractive for the closed loop system (5.4). Since $x^*(t) \in \mathcal{X}_p^* \subset \mathcal{X}_s^*$ implies $\delta(t) = 0$, $\mathcal{X}_p^*$ is invariant for the open-loop system (5.4a) – (5.4c) with $\delta(t) \equiv 0$. Thus, we obtain properties of the *uncoupled* systems by imposing conditions on the *coupled* systems, namely exponential convergence to a synchronized solution. The key observation is that couplings, since they are assumed to be of the diffusive type, vanish once the control objective, i.e., synchronization, is achieved. In fact, the couplings depend on synchronization errors, i.e., the synchronization errors are used as a controller input, and the control objective is to drive that input to zero. The situation is similar to the output regulation problem where the regulation error is used as a controller input, and the objective is to drive that regulation error to zero. The situation can be paraphrased as follows: since we want to achieve some property (i.e., synchronization) exponentially fast, there must be an invariant subspace (i.e., $\mathcal{X}_p^*$) on which the property is identically satisfied. Furthermore, since satisfaction of this property implies that the error signal driving the controller (i.e., $\delta(t) = (L(t) \otimes I_r)P^*x^*(t)$) vanishes, the set must be invariant for the open loop system, i.e., the uncoupled individual systems. The set $\mathcal{X}_p^*$ is non-trivial by assumption. In fact, $\mathcal{X}_p^* = \{0\}$ would imply that $y(t) \to 0$ independent of initial conditions $x^*(0) \in \mathbb{R}^\sigma$, i.e., $\mathcal{X}_p^* = \{0\}$ implies trivial synchronization.

Since $\mathcal{X}_p^*$ is invariant for the open loop system (5.4a) – (5.4c) with $\delta(t) \equiv 0$, we know that $x^*(t) = ((x_1^*(t))^T, \ldots, (x_N^*(t))^T)^T \in \mathcal{X}_p^*$ implies $\dot{x}_k^*(t) = A_k^* x_k^*(t)$ with $x_k^*(t) \in \mathcal{X}_{k,p}^* \triangleq \Xi_k \mathcal{X}_p^* \subset \mathbb{R}^{n_k+m_k}$, $k \in \mathbb{N}_N$, i.e., the individual systems evolve independently. The situation is illustrated in Figure 5.3 for a network of two systems. Since $\mathcal{X}_p^*$ is A^*-invariant and the individual systems are decoupled if $x^*(t) \in \mathcal{X}_p^*$, we know that $\mathcal{X}_{k,p}^*$ must be A_k^*-invariant subspaces for all $k \in \mathbb{N}_N$ and the dynamics of A_k^* restricted to $\mathcal{X}_{k,p}^*$ must be identical to the dynamics of A^* restricted to $\mathcal{X}_p^*$ for all $k \in \mathbb{N}_N$. Furthermore, since $x^*(t) \in \mathcal{X}_p^*$ implies $y(t) \in \mathcal{Y}_s$, i.e., $y_k(t) = y_j(t)$ for all $j, k \in \mathbb{N}_N$, we know that the systems (5.3) restricted to $\mathcal{X}_{k,p}^*$ are similar to the global system (5.4a) – (5.4c) with $\delta(t) \equiv 0$ restricted to $\mathcal{X}_p^*$ for all $k \in \mathbb{N}_N$. Thus in particular, the systems (5.3) restricted to $\mathcal{X}_{k,p}^*$ are equivalent for all $k \in \mathbb{N}_N$. This fact is expressed by conditions (5.5) in Theorem 5.4. The model of the systems (5.3) restricted to the subspaces $\mathcal{X}_{k,p}^*$ are given by a dynamical system

$$\dot{\xi}(t) = S\xi(t) \tag{5.7a}$$

$$\eta(t) = R\xi(t) \tag{5.7b}$$

with state vector $\xi(t) \in \mathbb{R}^\nu$ and output $\eta(t) \in \mathbb{R}^q$. By (5.6) we know that exponential synchronization in the network (5.3) implies that for any initial conditions $x_k^*(0) \in \mathbb{R}^{n_k+m_k}$, $k \in \mathbb{N}_N$ there exists an initial condition $\xi(0) \in \mathbb{R}^\nu$ such that $(y_k(t) - \eta(t)) \to 0$ exponentially fast as $t \to \infty$ where $\eta(t)$ is the output of (5.7). That is, the individual systems track the system (5.7) similar to systems tracking an exosystem in the output regulation problem. However, in synchronization problems, there is no true exosystem and (5.7) exists only as part of the network itself. Therefore we term system (5.7), where

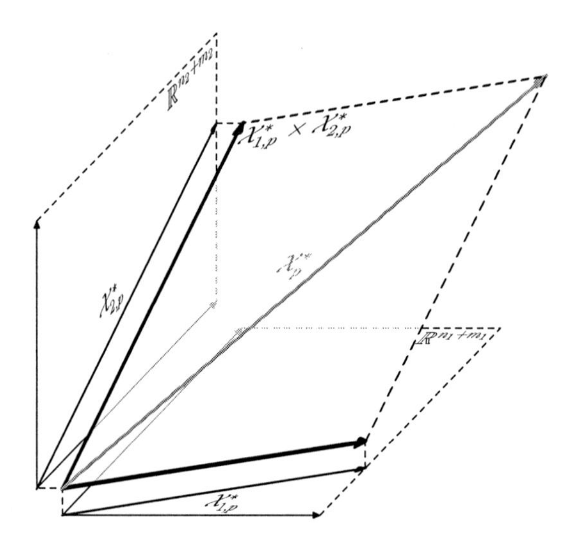

Figure 5.3: State spaces $\mathbb{R}^{n_1+m_1}$, $\mathbb{R}^{n_2+m_2}$ of two individual systems with dynamic coupling controllers with invariant subspaces $\mathcal{X}^*_{1,p}$ and $\mathcal{X}^*_{2,p}$ and invariant subspace $\mathcal{X}^*_p \subset \mathcal{X}^*_{1,p} \times \mathcal{X}^*_{2,p} \subset \mathbb{R}^\sigma$, $\sigma = \sum_{k=1}^{2}(n_k + m_k)$.

$S \in \mathbb{R}^{\nu\times\nu}$ and $R \in \mathbb{R}^{q\times\nu}$ satisfy conditions (5.5) of Theorem 5.4, *virtual exosystem* for the network (5.3).

Thus we can give a verbalized statement of Theorem 5.4 as follows:

> A necessary condition for solvability of the Linear Heterogeneous Output Synchronization Problem 5.1 is that the individual systems together with their dynamic coupling controllers all contain an *internal model* of some virtual exosystem (5.7)[1]. Furthermore, if synchronization is observed, the synchronous outputs are generated as outputs $\eta(t)$ of the virtual exosystem (5.7) for some initial condition $\xi(0) \in \mathbb{R}^\nu$ depending on initial conditions $x^*_k(0) \in \mathbb{R}^{n_k+m_k}$, $k \in \mathbb{N}_N$ and the communication topology.

The virtual exosystem (5.7) possesses the properties that $\sigma(S) \subset \overline{\mathbb{C}}_+$ and the pair (R, S) is observable. In fact, if $\sigma(S) \bigcap \mathbb{C}_- \neq \emptyset$, it is possible to decompose the virtual exosystem in an asymptotically stable system and a system with eigenvalues in the closed right-half complex plane. Since the asymptotically stable part asymptotically decays to zero, this part can be neglected. Observability of the pair (R, S) is a consequence of Assumption 5.3, more precisely detectability of the pair (C^*, A^*). If the pair (C^*, A^*) was not detectable, conditions (5.5) would still be satisfied but the pair (R, S) might not be observable. As a consequence, we would not be able to guarantee $\operatorname{rank}(\Psi_k) = \nu$ anymore. Thus the interpretation given above saying that any individual system contains an internal

[1] A system is said to contain an internal model of a virtual exosystem if its state space admits an invariant subset such that the system restricted to that subset is similar to the virtual exosystem.

model of the virtual exosystem (5.7) would not be valid anymore. Similar conditions and interpretations can however be obtained by considering only the observable part of the system (5.4).

To illustrate the implications of Theorem 5.4, we give two simple examples below.

Example 5.5. *Consider a network of $N = 2$ scalar systems modeled as $A_1 = a_1$, $B_1 = 1$, and $C_1 = 1$ and $A_2 = a_2$, $B_2 = 1$, and $C_2 = 1$ respectively, i.e., $n_k = p_k = q = 1$, $k = 1, 2$ with static coupling controllers $\zeta_k(t) = y_k(t)$, $u_k(t) = \kappa_k \delta_k(t)$, $\kappa_k \in \mathbb{R}$ for $k = 1, 2$, i.e., $m_k = 0$, $k = 1, 2$. Conditions (5.5) reduce to $A_k \Psi_k = \Psi_k S$ and $C_k \Psi_k = R$. With $C_1 = C_2 = 1$, we obtain $R = \Psi_1 = \Psi_2 \neq 0$. Since Ψ_k, $k = 1, 2$ are scalars, we obtain $a_k = S \geq 0$, i.e., $a_1 = a_2 \geq 0$ and thus the two systems are identical and not asymptotically stable.*

Necessity of this condition is verified as follows: The closed loop system reads

$$\begin{pmatrix} \dot{y}_1(t) \\ \dot{y}_2(t) \end{pmatrix} = \begin{pmatrix} a_1 + \kappa_1 w_{1,2}(t) & -\kappa_1 w_{1,2}(t) \\ -\kappa_2 w_{2,1}(t) & a_2 + \kappa_2 w_{2,1}(t) \end{pmatrix} \begin{pmatrix} y_1(t) \\ y_2(t) \end{pmatrix}.$$

Synchronization means $y_1(t) - y_2(t) \to 0$ as $t \to \infty$. Define $e(t) = y_1(t) - y_2(t)$ and $s(t) = y_1(t) + y_2(t)$. We have

$$\dot{e}(t) = \left(\frac{a_1 + a_2}{2} + \kappa_1 w_{1,2}(t) + \kappa_2 w_{2,1}(t) \right) e(t) + \left(\frac{a_1 - a_2}{2} \right) s(t).$$

Thus, $e(t) \to 0$ implies that either $a_1 = a_2$, i.e., the two systems are identical and conditions (5.5) are trivially satisfied or $s(t) \to 0$, i.e., trivial synchronization. If $a_1 = a_2$, the results from Chapter 3 can be applied to show under what assumptions (e.g., constant, connected communication graph or uniformly connected communication graph and $a_k \leq 0$, $k = 1, 2$) $a_1 = a_2$ is also sufficient for solvability of the synchronization problem.

Example 5.6. *As a second example consider again a network of $N = 2$ systems with dynamic coupling controllers modeled as*

$$A_1^* = \begin{pmatrix} 0 & 0 & 0 \\ 1 & -2 & 1 \\ 0 & -2 & 0 \end{pmatrix}, \quad B_1^* = \begin{pmatrix} 1 \\ 0 \\ -1 \end{pmatrix}, \quad C_1^* = \begin{pmatrix} 2 & -1 & 1 \end{pmatrix},$$
$$A_2^* = \begin{pmatrix} 0 & 1 \\ 0 & -1 \end{pmatrix}, \quad B_2^* = \begin{pmatrix} 1 \\ 0 \end{pmatrix}, \quad C_2^* = \begin{pmatrix} 1 & -1 \end{pmatrix}.$$

The conditions of Theorem 5.4 are satisfied with $\Psi_1 = (1, 0, -1)^T$, $\Psi_2 = (1, 0)^T$, $S = 0$, and $R = 1$. The pair (R, S) is trivially observable and $\operatorname{rank}(\Psi_1) = \operatorname{rank}(\Psi_2) = 1$. In fact, minimal realizations of both individual systems with dynamic coupling controllers are given as $\dot{y}_k(t) = u_k(t)$, $k = 1, 2$. Thus, we can again apply methods from Chapter 3 to synchronize the network. For instance $u_k(t) = \delta_k(t)$, $k = 1, 2$ yields synchronization for any uniformly connected communication graph by Theorem 3.20.

Example 5.6 suggests a general procedure to synchronize heterogeneous networks by designing dynamic coupling controllers such that the input-output behavior for all systems is the same and then apply the results from Chapters 3 and 4 to the controllable and observable part of the individual systems with their dynamic coupling controller. Theorem 5.4 conveys that, if the pairs (C_k^*, A_k^*), $k \in \mathbb{N}_N$ are detectable, a necessary condition for this to be possible is given by (5.5).

As mentioned earlier, these conditions for solvability of the Linear Heterogeneous Output Synchronization Problem 5.1 given in Theorem 5.4 have an important drawback. Namely, they depend on the solution to the problem, i.e., they are implicit conditions. We are however interested in explicit conditions that depend on the problem data only, i.e., the description of the system models (5.3). Such conditions are derived below.

Explicit Internal Model Principle

The explicit conditions are in fact obtained as a corollary of the implicit conditions given in Theorem 5.4. To state that corollary, we first need an intermediate result relating conditions (5.5) in Theorem 5.4 to conditions that depend only on the problem data.

Lemma 5.7. *Let $k \in \mathbb{N}_N$ fixed. Let $A_k \in \mathbb{R}^{n_k \times n_k}$, $B_k \in \mathbb{R}^{n_k \times p_k}$, $C_k \in \mathbb{R}^{q \times n_k}$, $S \in \mathbb{R}^{\nu \times \nu}$, and $R \in \mathbb{R}^{q \times \nu}$.*

There exist matrices $E_k \in \mathbb{R}^{m_k \times m_k}$, $F_k \in \mathbb{R}^{m_k \times r}$, $M_k \in \mathbb{R}^{m_k \times q}$, $G_k \in \mathbb{R}^{p_k \times m_k}$, $H_k \in \mathbb{R}^{p_k \times r}$, $O_k \in \mathbb{R}^{p_k \times q}$, $P_k \in \mathbb{R}^{r \times m_k}$, $Q \in \mathbb{R}^{r \times q}$, $\Psi_k \in \mathbb{R}^{(n_k+m_k) \times \nu}$ such that conditions (5.5) are satisfied (with $A_k^* \in \mathbb{R}^{(n_k+m_k)\times(n_k+m_k)}$ and $C_k^* \in \mathbb{R}^{q \times (n_k+m_k)}$ as defined before) if and only if there exist matrices $\Pi_k \in \mathbb{R}^{n_k \times \nu}$ and $\Lambda_k \in \mathbb{R}^{p_k \times \nu}$ that solve

$$A_k \Pi_k + B_k \Lambda_k = \Pi_k S, \tag{5.8a}$$

$$C_k \Pi_k = R. \tag{5.8b}$$

Proof. Assume $\Psi_k \in \mathbb{R}^{(n_k+m_k)\times\nu}$ solves (5.5). Let $\Psi_k = (\Pi_k^T, \Sigma_k^T)^T$ with $\Pi_k \in \mathbb{R}^{n_k \times \nu}$ and $\Sigma_k \in \mathbb{R}^{m_k \times \nu}$. Substituting the definitions of A_k^* and B_k^* into (5.5) yields

$$\begin{aligned} A_k \Pi_k + B_k(O_k C_k \Pi_k + G_k \Sigma_k) &= \Pi_k S, \\ M_k C_k \Pi_k + E_k \Sigma_k &= \Sigma_k S, \\ C_k \Pi_k &= R. \end{aligned}$$

Necessity is shown by defining $\Lambda_k \triangleq O_k C_k \Pi_k + G_k \Sigma_k = O_k R + G_k \Sigma_k$. To prove sufficiency, we thus need to show that the equations

$$\begin{aligned} E_k \Sigma_k &= \Sigma_k S - M_k R, \\ G_k \Sigma_k &= \Lambda_k - O_k R \end{aligned}$$

admit a solution E_k, G_k, M_k, O_k, and Σ_k for given matrices S, R, and Λ_k. One such solution is given as $E_k = S$, $G_k = \Lambda_k$, $M_k = 0$, $O_k = 0$ and $\Sigma_k = I_\nu$. □

By virtue of the above lemma, it is easy to obtain an explicit solvability condition for the Linear Heterogeneous Output Synchronization Problem 5.1. Before doing so, we want to shortly comment on the sufficiency part of the above proof, which has been carried out constructively. Given a solution $\Pi_k \in \mathbb{R}^{n_k \times \nu}$, $\Lambda_k \in \mathbb{R}^{p_k \times \nu}$ to (5.8), the dynamic couplings (5.2) chosen as $E_k = S$, $G_k = \Lambda_k$, $M_k = 0$, $O_k = 0$ are such that $\Psi_k = (\Pi_k^T, I_\nu^T)^T$ solves the implicit conditions (5.5a). The matrices F_k, H_k, P_k and Q of the dynamic couplings are left unspecified. If we set them to be equal to zero, we obtain dynamic couplings that act in a purely feedforward manner on the corresponding individual system. This feedforward control ensures that the set $\{(x_k, z_k) \in \mathbb{R}^{n_k} \times \mathbb{R}^{m_k} | x_k = \Pi_k z_k\}$ is invariant for (5.3) and the dynamics restricted to this set are those of the virtual exosystem (5.7a). Of course, the dynamic couplings proposed above will generally not solve the Linear Heterogeneous Output Synchronization Problem 5.1 on their own but they will need to be extended by a stabilizing part using the remaining degrees of freedom.

Using the above result, the explicit solvability condition is stated in the corollary below.

Corollary 5.8. *Let $\mathcal{G}(t) = \{\mathcal{V}, \mathcal{E}(t), W(t)\}$ be some communication graph with $|\mathcal{V}| = N > 1$ and Laplacian matrix $L(t)$. Let $A_k \in \mathbb{R}^{n_k \times n_k}$, $B_k \in \mathbb{R}^{n_k \times p_k}$, and $C_k \in \mathbb{R}^{q \times n_k}$ for $k \in \mathbb{N}_N$.*

If there exist matrices $E_k \in \mathbb{R}^{m_k \times m_k}$, $F_k \in \mathbb{R}^{m_k \times r}$, $M_k \in \mathbb{R}^{m_k \times q}$, $G_k \in \mathbb{R}^{p_k \times m_k}$, $H_k \in \mathbb{R}^{p_k \times r}$, $O_k \in \mathbb{R}^{p_k \times q}$, $P_k \in \mathbb{R}^{r \times m_k}$ for $k \in \mathbb{N}_N$, and $Q \in \mathbb{R}^{r \times q}$ that solve the Linear Heterogeneous Output Synchronization Problem 5.1 while satisfying Assumption 5.3, then there exists a positive integer $\nu \in \mathbb{N}$ and matrices $\Pi_k \in \mathbb{R}^{n_k \times \nu}$, $\Lambda_k \in \mathbb{R}^{p \times \nu}$ for $k \in \mathbb{N}_N$, $S \in \mathbb{R}^{\nu \times \nu}$, and $R \in \mathbb{R}^{q \times \nu}$ such that $\sigma(S) \subset \overline{\mathbb{C}}_+$, the pair (R, S) is observable, and (5.8) holds for all $k \in \mathbb{N}_N$. Furthermore, there exists a vector $\xi_0 \in \mathbb{R}^\nu$, depending on initial conditions $x_k(0) \in \mathbb{R}^{n_k}$, $z_k(0) \in \mathbb{R}^{m_k}$, $k \in \mathbb{N}_N$ and the communication graph $\mathcal{G}(t)$, and constants $M, \mu \in \mathbb{R}_+$ such that (5.6) holds for all $k \in \mathbb{N}_N$.

Proof. Since solvability of (5.5) in the unknowns $\Psi_k \in \mathbb{R}^{(n_k+m_k) \times \nu}$ implies solvability of (5.8) in the unknowns $\Pi_k \in \mathbb{R}^{n_k \times \nu}$ and $\Lambda_k \in \mathbb{R}^{p_k \times \nu}$ for all $k \in \mathbb{N}_N$ by Lemma 5.7, the corollary follows from Theorem 5.4. □

We showed in Lemma 5.7, that solvability of conditions (5.5) given in Theorem 5.4 is equivalent to solvability of conditions (5.8) involving the individual system matrices $A_k \in \mathbb{R}^{n_k \times n_k}$ and $C_k \in \mathbb{R}^{q \times n_k}$, the virtual exosystem matrices $S \in \mathbb{R}^{\nu \times \nu}$ and $R \in \mathbb{R}^{q \times \nu}$, and the unknowns $\Pi_k \in \mathbb{R}^{n_k \times \nu}$ and $\Lambda_k \in \mathbb{R}^{p_k \times \nu}$. We do not know yet whether solvability of (5.8) implies solvability of (5.5) under the additional constraints imposed by Assumption 5.3. Though we know that the conditions of Theorem 5.4 imply the conditions of Corollary 5.8 and the conditions of Corollary 5.8 imply the conditions of Theorem 5.4 except for the constraints imposed by Assumption 5.3. In subsequent chapters of this thesis, we will show that the conditions are actually equivalent under suitable assumptions.

The conditions (5.8) correspond to the well-known *Francis equations* for the tracking problem in the framework of linear systems (see Francis, 1977, Francis and Wonham, 1975, Trentelman et al., 2001). Conditions (5.8a) implies that the system

$$\begin{pmatrix} \dot{x}_k(t) \\ \dot{\xi}(t) \end{pmatrix} = \begin{pmatrix} A_k & 0 \\ 0 & S \end{pmatrix} \begin{pmatrix} x_k(t) \\ \xi(t) \end{pmatrix} + \begin{pmatrix} B_k \\ 0 \end{pmatrix} u_k(t)$$

with $x_k(t) \in \mathbb{R}^{n_k}$, $\xi(t) \in \mathbb{R}^\nu$, and $u_k(t) \in \mathbb{R}^{p_k}$ admits a controlled invariant subspace $\{(x_k, \xi) \in \mathbb{R}^{n_k} \times \mathbb{R}^\nu : x_k = \Pi_k \xi\} \subset \mathbb{R}^{n_k} \times \mathbb{R}^\nu$. The subspace is rendered invariant using the control $u_k(t) = \Lambda_k \xi(t)$. Condition (5.8b) implies that on the above subspace, i.e., for $x_k(t) = \Pi \xi(t)$, $y_k(t) = C_k x_k(t) = R\xi(t) = \eta(t)$, i.e., the system outputs $y_k(t)$, $k \in \mathbb{N}_N$ are identical to the output $\eta(t)$ of the virtual exosystem. Recall that the condition (5.5a) in

Theorem 5.4 implies that the controlled system (5.3) contains an internal model of the virtual exosystem, i.e., the system

$$\begin{pmatrix} \dot{x}_k^*(t) \\ \dot{\xi}(t) \end{pmatrix} = \begin{pmatrix} A_k^* & 0 \\ 0 & S \end{pmatrix} \begin{pmatrix} x_k^*(t) \\ \xi(t) \end{pmatrix}$$

with $x_k^*(t) \in \mathbb{R}^{n_k+m_k}$ and $\xi(t) \in \mathbb{R}^\nu$ admits an invariant subspace $\{(x_k^*, \xi) \in \mathbb{R}^{(n_k+m_k)} \times \mathbb{R}^\nu : x_k^* = \Psi_k \xi\} \subset \mathbb{R}^{(n_k+m_k)} \times \mathbb{R}^\nu$ while condition (5.5b) implies that on that subspace $y_k(t) = C_k^* x_k^*(t) = R\xi(t) = \eta(t)$. It appears natural that imposing existence of an invariant subspace with certain properties for a controlled system translates to the requirement that a controlled invariant subspace possessing the same properties exists for the open-loop system. That is exactly what the conditions in Corollary 5.8 are.

A comparison between the implicit conditions from Theorem 5.4 and the explicit conditions from Corollary 5.8 yields an interesting (though not at all surprising) consequence for synchronization of heterogeneous networks with static diffusive couplings stated in the remark below.

Remark 5.9. *In the special case of static diffusive couplings, we have* $m_k = 0$ *and* $A_k^* = A_k$, $C_k^* = C_k$ *for all* $k \in \mathbb{N}_N$. *Thus, solvability of the Linear Heterogeneous Output Synchronization Problem with static diffusive couplings implies that* (5.8) *has a solution* $\Pi_k \in \mathbb{R}^{n_k \times \nu}$, $\Lambda_k \in \mathbb{R}^{p_k \times \nu}$ *with* $\Lambda_k = 0$ *and* $\operatorname{rank}(\Pi_k) = \nu$ *for all* $k \in \mathbb{N}_N$. *This is of course much more restrictive than requiring existence of any solution* $\Pi_k \in \mathbb{R}^{n_k \times \nu}$, $\Lambda_k \in \mathbb{R}^{p_k \times \nu}$ *to* (5.8). *As a consequence, dynamic couplings are in general a necessary precondition for solvability of the Linear Heterogeneous Output Synchronization Problem 5.1. This observation is in perfect alignment with the idea promoted in the present thesis, that dynamic couplings are needed to allow for increased system and topological complexity in consensus and synchronization problems.*

The Virtual Exosystem

In classical tracking problems, the type of trajectories that are to be tracked, and thus the exosystem generating those trajectories, is known a priori. The problem to solve is to find a tracking controller for a given exosystem. The internal model conditions presented in Theorem 5.4 and Corollary 5.8 merely require existence of some virtual exosystem an internal model of which is embedded in the individual system models in an appropriate sense. The relevance of this virtual exosystem is highlighted in condition (5.6) in Theorem 5.4. Namely, the virtual exosystem (5.7) determines the possible synchronous trajectories of the network. Thus, by choosing a particular virtual exosystem, we determine the asymptotic behavior of the network if synchronization can be achieved. Therefore, we next address the question what conditions need to be satisfied by the virtual exosystem such that the conditions of Theorem 5.4 or Corollary 5.8 may be satisfied.

Since the Francis equations (5.8) are well known from output regulation problems, we can adopt the solvability conditions that were proved in Hautus (1983) (see also Knobloch et al., 1993, Trentelman et al., 2001). To this end, we repeat the following Lemma without proof:

Lemma 5.10. *Let $V_k \in \mathbb{R}^{\overline{n}\times\overline{m}}$, $k \in \mathbb{N}_r$ and $W \in \mathbb{R}^{\overline{p}\times\overline{p}}$ be constant matrices and let $q_k(\cdot) : \mathbb{R} \to \mathbb{R}$, $k \in \mathbb{N}_r$ be polynomials. The equation*

$$\sum_{k=1}^{r} V_k X q_k(W) = Z \tag{5.9}$$

admits a solution $X \in \mathbb{R}^{\overline{m}\times\overline{p}}$ for any $Z \in \mathbb{R}^{\overline{n}\times\overline{p}}$ if and only if

$$\operatorname{rank}\left(\sum_{k=1}^{r} q_k(s)V_k\right) = \overline{n} \quad \textit{for all} \quad s \in \sigma(W).$$

With the help of Lemma 5.10, we can state and proof a well known result which gives sufficient conditions for solvability of conditions (5.8).

Lemma 5.11. *Let $A_k \in \mathbb{R}^{n_k\times n_k}$, $B_k \in \mathbb{R}^{n_k\times p_k}$, $C_k \in \mathbb{R}^{q\times n_k}$, $S \in \mathbb{R}^{\nu\times\nu}$ and $R \in \mathbb{R}^{q\times\nu}$ be constant matrices.*

If

$$\operatorname{rank}\begin{pmatrix} sI - A_k & -B_k \\ -C_k & 0 \end{pmatrix} = n_k + q \quad \textit{for all} \quad s \in \sigma(S), \tag{5.10}$$

then conditions (5.8) *admit a solution $\Pi_k \in \mathbb{R}^{n_k\times\nu}$, $\Lambda_k \in \mathbb{R}^{p_k\times\nu}$.*

Proof. In order to apply Lemma 5.10 to conditions (5.8) given in Corollary 5.8, we rewrite these conditions as

$$\begin{pmatrix} A_k & B_k \\ C_k & 0 \end{pmatrix}\begin{pmatrix} \Pi_k \\ \Lambda_k \end{pmatrix} I_\nu + \begin{pmatrix} -I_{n_k} & 0 \\ 0 & 0 \end{pmatrix}\begin{pmatrix} \Pi_k \\ \Lambda_k \end{pmatrix} S = \begin{pmatrix} 0 \\ -R \end{pmatrix}.$$

This equality is of the form (5.9), and we can identify the matrices

$$V_1 = \begin{pmatrix} A_k & B_k \\ C_k & 0 \end{pmatrix}, \quad V_2 = \begin{pmatrix} -I_{n_k} & 0 \\ 0 & 0 \end{pmatrix}, \quad X = \begin{pmatrix} \Pi_k \\ \Lambda_k \end{pmatrix}, \quad Z = \begin{pmatrix} 0 \\ -R \end{pmatrix},$$

and $W = S$, and the polynomials $q_1(s) = 1$ and $q_2(s) = s$. Using Lemma 5.10 we thus know that conditions (5.8) are solvable for $\Pi_k \in \mathbb{R}^{n_k\times\nu}$ and $\Lambda_k \in \mathbb{R}^{p_k\times\nu}$ if (5.10) holds. □

Condition (5.10) in Lemma 5.11 can only be satisfied if the individual systems have at least as many inputs as outputs, i.e., $p_k \geq q$, $k \in \mathbb{N}_N$. In that case, condition (5.10) can be interpreted as a non-resonance condition in the sense that no zero of an individual system must be a pole of the virtual exosystem. Note that condition (5.10) is sufficient for solvability of (5.8) but generally not necessary. Necessary and sufficient conditions exist (see Trentelman et al., 2001) in terms of equating transfer matrices of different systems. However, since condition (5.10) does not impose an important restriction on the choice of S, we stick to the conditions of Lemma 5.11.

To conclude the section on the linear internal model principle, we give a short summary. The point of departure has been a network of N non-identical LTI systems and we were interested in necessary conditions for existence of dynamic coupling controllers, subject to the constraint of exchanging only relative information over the network, that guarantee non-trivial output-synchronization. It turned out, that systems can synchronize asymptotically only if the closed loop system possesses an invariant subspace on which the outputs of all individual systems are identical. This translates into the requirement that every individual system together with its dynamic coupling controller embeds an internal model of some virtual exosystem, which determines the behavior of the network once all individuals are synchronized. Links to the well-known output regulation problem for linear systems have been established at several places. In particular, the necessary condition for the individual system has been given in terms of linear matrix equations, which are known as Francis equations in the context of output regulation problems. In the last subsection, we stressed that the virtual exosystem can be considered as a degree of freedom of the design, i.e., one can choose to what type of trajectories the individual systems shall synchronize. This choice is however constraint by properties of the individual systems.

5.2 The Nonlinear Case

In what follows, we will investigate to what extent the ideas presented for synchronization in networks of non-identical linear systems (5.1) possess a nonlinear analogue. We will first present a nonlinear version of the heterogeneous output synchronization problem and derive necessary conditions for solvability of that problem – under appropriate assumptions – subsequently. The results presented in this chapter are largely based upon Wieland and Allgöwer (2009b).

5.2.1 Problem Setup

The network under consideration is the nonlinear generalization of the network (5.1) considered in Section 5.2.1, i.e., we consider a group of N systems modeled by nonlinear ODEs of the form

$$\dot{x}_k(t) = f_k(x_k(t), u_k(t)) \tag{5.11a}$$

$$y_k(t) = h_k(x_k(t)) \tag{5.11b}$$

where $x_k(t) \in \mathbb{R}^{n_k}$ is the state vector, $u_k(t) \in \mathbb{R}^{p_k}$ is the input vector and $y_k(t) \subset \mathbb{R}^q$ is the output vector for $k \in \mathbb{N}_N$. As before, we do not require that $n_i = n_j$ or $p_i = p_j$ for $i, j \in \mathbb{N}_N, i \neq j$, i.e., the dynamics of the individual systems in the group, modeled by the maps $f_k : \mathbb{R}^{n_k} \times \mathbb{R}^{p_k} \to \mathbb{R}^{n_k}$ and $h_k : \mathbb{R}^{n_k} \to \mathbb{R}^q$, may be different including state and input dimensions. Since the problem under consideration is to synchronize the outputs of the systems (5.11), the outputs are constrained to have the same dimension. In order to guarantee existence and uniqueness of solutions, we assume that the maps f_k and h_k are locally Lipschitz in their arguments.

The nonlinear analogue to couplings (5.2) from Section 5.1.1 are given as

$$\dot{z}_k(t) = \phi_k(z_k(t), \delta_k(t), y_k(t)) \tag{5.12a}$$
$$u_k(t) = \alpha_k(z_k(t), \delta_k(t), y_k(t)) \tag{5.12b}$$
$$\zeta_k(t) = \beta_k(z_k(t), y_k(t)) \tag{5.12c}$$
$$\delta_k(t) = \sum_{j=1}^{N} w_{k,j}(t) d(\zeta_k(t), \zeta_j(t)) \tag{5.12d}$$

with coupling state $z_k(t) \in \mathbb{R}^{m_k}$ and virtual outputs $\zeta_k(t) \in \mathbb{R}^r$ for $k \in \mathbb{N}_N$. Couplings (5.12) represent a general nonlinear dynamic controller driven by the system outputs $y_k(t)$ and the relative virtual output signal $\delta_k(t)$. The outputs of the controller are the system input $u_k(t)$ and the virtual output $\zeta_k(t)$ obtained as a nonlinear function of the system outputs $y_k(t)$ and the controller states $z_k(t)$. Again, the values $w_{k,j}(t) \in \overline{\mathbb{R}}_+$, $j, k \in \mathbb{N}_N$ are the elements of the adjacency matrix $W(t) = [w_{k,j}(t)] \in \mathbb{R}^{N \times N}$ of some communication graph $\mathcal{G}(t) = \{\mathcal{V}, \mathcal{E}(t), W(t)\}$. The situation is thus identical to the situation depicted in the block diagrams in Figure 5.1 for the linear case. The maps $\phi_k : \mathbb{R}^{m_k} \times \mathbb{R}^r \times \mathbb{R}^q \to \mathbb{R}^{m_k}$, $\alpha_k : \mathbb{R}^{m_k} \times \mathbb{R}^r \times \mathbb{R}^q \to \mathbb{R}^{p_k}$, and $\beta_k : \mathbb{R}^{m_k} \times \mathbb{R}^q \to \mathbb{R}^r$ are assumed to be locally Lipschitz in their arguments to guarantee existence and uniqueness of solutions. The map $d : \mathbb{R}^r \times \mathbb{R}^r \to \mathbb{R}^r$ models the *relative values* of the virtual outputs ζ_k of the dynamic coupling controllers (in the linear case, $d(\zeta_k(t), \zeta_j(t)) = \zeta_k(t) - \zeta_j(t)$), i.e., (5.12d) is a nonlinear generalization of the diffusive couplings considered so far. Since we are interested in necessary conditions for solvability of the heterogeneous output synchronization problem, we do not need to define precisely the meaning of *relative value* and the properties this imposes for the map d. We will merely assume that d is locally Lipschitz in its arguments and $d(\zeta, \zeta) = 0$ for all $\zeta \in \mathbb{R}^r$.

We define the closed loop of systems (5.11) together with the couplings (5.12) as a single system

$$\dot{x}_k^*(t) = f_k^*(x_k^*(t), \delta_k(t)) \tag{5.13a}$$
$$y_k(t) = h_k^*(x_k^*(t)) \tag{5.13b}$$
$$\zeta_k(t) = \beta_k^*(x_k^*(t)) \tag{5.13c}$$

for $k \in \mathbb{N}_N$ with extended state vector $x_k^*(t) = (x_k(t)^T, z_k(t)^T)^T \in \mathbb{R}^{n_k+m_k}$ and maps

$$f_k^*(x_k^*(t), \delta_k(t)) \triangleq \begin{pmatrix} f_k\big(x_k(t), \alpha_k\big(z_k(t), \delta_k(t), h_k(x_k(t))\big)\big) \\ \phi_k\big(z_k(t), \delta_k(t), h_k(x_k(t))\big) \end{pmatrix},$$
$$h_k^*(x_k^*(t)) \triangleq h_k(x_k(t)), \quad \beta_k^*(x_k^*(t)) \triangleq \beta_k\big(z_k(t), h_k(x_k(t))\big).$$

Note that the maps $f_k^* : \mathbb{R}^{n_k+m_k} \times \mathbb{R}^r \to \mathbb{R}^{n_k+m_k}$, $h_k^* : \mathbb{R}^{n_k+m_k} \to \mathbb{R}^q$, and $\beta_k^* : \mathbb{R}^{n_k+m_k} \to \mathbb{R}^r$ are locally Lipschitz in their arguments if the original system and controller maps are locally Lipschitz. We define the global system obtained from (5.13) with couplings (5.12d) as

$$\dot{x}^*(t) = f^*(x^*(t), \delta(t)) \tag{5.14a}$$
$$y(t) = c^*(x^*(t)) \tag{5.14b}$$
$$\zeta(t) = \beta^*(x^*(t)) \tag{5.14c}$$
$$\delta(t) = D(t, \zeta(t)) \tag{5.14d}$$

with stacked state vector $x^*(t) = (x_1^*(t)^T, \dots, x_N^*(t)^T)^T \in \mathbb{R}^\sigma$ where $\sigma = \sum_{k=1}^N (n_k + m_k)$, stacked system output vector $y(t) = (y_1(t)^T, \dots, y_N(t)^T)^T \in \mathbb{R}^{Nq}$, and stacked virtual output vector $\zeta(t) = (\zeta_1(t)^T, \dots, \zeta_N(t)^T)^T \in \mathbb{R}^{Nr}$. The maps $f^* : \mathbb{R}^\sigma \times \mathbb{R}^{Nr} \to \mathbb{R}^\sigma$, $h^* : \mathbb{R}^\sigma \to \mathbb{R}^{Nq}$, and $\beta^* : \mathbb{R}^\sigma \to \mathbb{R}^{Nr}$ are obtained by stacking the maps for the individual systems defined above. The map $D(t, \zeta(t)) : \mathbb{R} \times \mathbb{R}^{Nr} \to \mathbb{R}^{Nr}$ summarizes the couplings in the network. If $d(\zeta_k, \zeta_j)$ is locally Lipschitz in its arguments and $w_{k,j}(t)$ is piecewise continuous for all $j, k \in \mathbb{N}_N$, $D(t, \zeta)$ will be piecewise continuous in t and locally Lipschitz in ζ. Furthermore, if $d(\zeta, \zeta) = 0$ for all $\zeta \in \mathbb{R}^r$, we have $D(t, 1_N \otimes \zeta) = 0$ for all $t \in \mathbb{R}$ and all $\zeta \in \mathbb{R}^r$. The latter property is a consequence of the fact that $D(t, \zeta(t))$ depends only on relative values of pairs of virtual outputs $\zeta_k(t)$, $k \in \mathbb{N}_N$.

With those definitions in place, we are ready to state the nonlinear analogue to the Linear Heterogeneous Output Synchronization Problem 5.1 as follows:

Problem 5.12 (Nonlinear Heterogeneous Output Synchronization). *Let the N individual systems* (5.11) *be modeled by locally Lipschitz maps $f_k : \mathbb{R}^{n_k} \times \mathbb{R}^{p_k} \to \mathbb{R}^{n_k}$ and $h_k : \mathbb{R}^{n_k} \to \mathbb{R}^q$ for $k \in \mathbb{N}_N$. Let $\mathcal{X}_k \subset \mathbb{R}^{n_k}$, $k \in \mathbb{N}_N$ be closed sets. Let the communication topology in the couplings* (5.12d) *be defined by some communication graph $\mathcal{G}(t) = \{\mathcal{V}, \mathcal{E}(t), W(t)\}$ with $|\mathcal{V}| = N$ and let $d : \mathbb{R}^r \times \mathbb{R}^r \to \mathbb{R}^r$ be a locally Lipschitz map satisfying $d(\zeta, \zeta) = 0$ for all $\zeta \in \mathbb{R}^r$.*

Find, if possible, locally Lipschitz maps $\phi_k : \mathbb{R}^{m_k} \times \mathbb{R}^r \times \mathbb{R}^q \to \mathbb{R}^{m_k}$, $\alpha_k : \mathbb{R}^{m_k} \times \mathbb{R}^r \times \mathbb{R}^q \to \mathbb{R}^{p_k}$, $\beta_k : \mathbb{R}^{m_k} \times \mathbb{R}^q \to \mathbb{R}^r$, and closed sets $\mathcal{Z}_k \subset \mathbb{R}^{m_k}$ such that the closed loop system (5.14) *satisfies the following properties:*

(P1) The set $\mathcal{X}^* = \mathcal{X}_1 \times \mathcal{Z}_1 \times \cdots \times \mathcal{X}_N \times \mathcal{Z}_N \subset \mathbb{R}^\sigma$ is positively invariant for (5.14)*, i.e., solutions $x^*(t_0 + t)$ of* (5.14) *with $x(t_0) \in \mathcal{X}^*$ for any $t_0 \in \mathbb{R}$ exist and remain in $\mathcal{X}^*$ for all $t \geq 0$.*

(P2) Solutions $x^*(t_0 + t)$ of (5.14) *with $x^*(t_0) \in \mathcal{X}^*$ are ultimately bounded uniformly in initial time t_0, i.e., there exists a bounded set $\mathcal{B}^* \subset \mathcal{X}^*$ with the property that for any compact subset $\mathcal{X}_0^* \subset \mathcal{X}^*$, there exists some time $T \in \mathbb{R}_+$ such that all solutions of* (5.14) *with initial data $(t_0, x(t_0)) \in \mathbb{R} \times \mathcal{X}_0^*$ satisfy $x(t_0 + t) \in \mathcal{B}^*$ for all $t \geq T$.*

(P3) The system outputs and the virtual outputs uniformly synchronize, i.e., $(y_k(t_0+t) - y_j(t_0+t)) \to 0$ as $t \to \infty$ and $(\zeta_k(t_0+t) - \zeta_j(t_0+t)) \to 0$ as $t \to \infty$ for all $j, k \in \mathbb{N}_N$ uniformly in initial data $(t_0, x^*(t_0)) \in \mathbb{R} \times \mathcal{X}_0^*$ for any compact subset $\mathcal{X}_0^* \subset \mathcal{X}^*$.

As before, we are not interested in solutions to the Nonlinear Heterogeneous Output Synchronization Problem 5.12 in first place, but we aim at answering the question what can be deduced about the individual systems and their dynamic coupling controllers if we know that there exists a solution to Problem 5.12. That is, we are again interested in *necessary conditions* for solvability of Problem 5.12 imposed on the individual systems (5.11).

However, before deriving those conditions, we shortly comment on the Properties (P1) – (P3) given in the problem statement and compare Problem 5.12 to Problem 5.1. In Problem 5.1 we aimed at global exponential synchronization among linear systems while Problem 5.12 aims at synchronization for initial conditions taken in some closed set $\mathcal{X}^* \subset \mathbb{R}^\sigma$. In order for the problem to be well defined, we need the set $\mathcal{X}^*$ to be positively invariant which is stated in Property (P1). Property (P2) means, that we are not interested in synchronized solutions that grow unbounded over time. This is certainly a reasonable requirement in many instances. We did not have such a requirement for Problem 5.1 because it actually makes no difference in the resulting conditions in the linear case. The simple reason for this is that invariant sets for linear systems are subspaces which implies that it is sufficient to know the behavior of the system near the origin to know its behavior everywhere. This is certainly not true for nonlinear systems, where it is generally hard to characterize the asymptotic behavior if solutions grow unbounded. The various uniformity assumptions in Properties (P2) and (P3) ensure that convergence times can grow unbounded only if initial conditions grow unbounded, i.e., it excludes singular points in the state space. Since convergence is always uniform for linear systems similar requirements have not been included for the linear case in Problem 5.1.

5.2.2 The Internal Model Principle for Synchronization among Nonlinear Systems

In Section 5.1.2, we showed that solvability of Problem 5.1 implies existence of a nontrivial persistent invariant synchronous subspace for the global system. Since we are considering nonlinear systems here, linear subspaces are clearly not suited for deriving solvability conditions of Problem 5.12. However, there may still be some *invariant sets* which take over the role played by invariant linear subspaces in Section 5.1. In fact, in the linear problem we showed that synchronization implies that all solutions converge to an invariant set for the global closed loop system (a linear subspace for the linear problem), on which the outputs are identically synchronous and solutions $x^*(t)$ starting in that set are persistent, i.e., do not converge to the origin (unless $x^*(t) \equiv 0$). It can thus be argued that this invariant set characterizes the steady state locus of the linear network considered in Section 5.1. The dynamics restricted to this set, i.e., the dynamics of the virtual exosystem (5.7), characterizes the steady state behavior of the network.

It will turn out that, under appropriate assumptions, similar arguments are valid for the nonlinear case. We will prove that, in order to synchronize the network (5.11), (5.12), the global closed loop system must admit a steady state locus, i.e., an invariant set which attracts all solutions and on which solutions persist, such that all outputs are synchronous once the steady state is reached. A large portion of the subsequent discussion will be dedicated to characterizing adequately the invariant sets that define the steady state locus of system (5.14). The notion of ω-limit sets, explained in some detail in Appendix B.2, will be essential for that purpose (see also Byrnes and Isidori, 2003, Isidori and Byrnes, 2008).

Implicit Internal Model Principle

As in the linear case, we will first derive implicit solvability conditions for Problem 5.12, i.e., conditions that depend on the problem data and the solution of the problem. In a second step, we will try to remove the dependence on the solution to obtain necessary solvability conditions for Problem 5.12 that exclusively depend on the problem data.

The implicit version of the solvability conditions for the nonlinear Heterogeneous Output Synchronization Problem are derived below. After some preliminary results, we will proceed in two steps. First, we prove that solvability of Problem 5.12 implies existence of a set characterizing the steady state locus for system (5.14) on which the outputs are identically synchronous. In a second step, we characterize the steady state behavior by giving an internal model interpretation.

We first give some results concerning the set $\mathcal{B}^* \subset \mathcal{X}^*$ used in Property (P2) of the problem statement to characterize ultimate boundedness of solutions. Clearly, the set $\mathcal{B}^*$ is not unique. For, given any bounded set $\mathcal{B}^*$ which ultimately bounds solutions of (5.14), any bounded set $\hat{\mathcal{B}}^* \supset \mathcal{B}^*$ will also ultimately bound solutions of (5.14). However, we have the following result adapted from Isidori and Byrnes (2008, Lemma 5):

Lemma 5.13. *Given a system $\dot{x}(t) = f(t, x(t))$ where $f : \mathbb{R} \times \mathbb{R}^n \to \mathbb{R}^n$ is piecewise continuous in the first argument and locally Lipschitz in the second argument. Suppose $\mathcal{X} \subset \mathbb{R}^n$ is a closed set such that solutions $x(t_0 + t)$ with initial data $(t_0, x(t_0)) \subset \mathbb{R} \times \mathcal{X}$ exist and remain in $\mathcal{X}$ for all $t \in \overline{\mathbb{R}}_+$ and are ultimately bounded uniformly in initial time $t_0 \in \mathbb{R}$, i.e., there exists a bounded set $\mathcal{B}$ with the property that for any compact subset $\mathcal{X}_0 \subset \mathcal{X}$ there exists a time $T \in \mathbb{R}_+$ such that solutions $x(t_0 + t)$ with initial data $(t_0, x(t_0)) \in \mathbb{R} \times \mathcal{X}_0$ satisfy $x(t_0 + t) \in \mathcal{B}$ for all $t \geq T$.*

Then $\omega(\mathbb{R} \times \hat{\mathcal{B}}) \subset \omega(\mathbb{R} \times \mathcal{B})$ for any bounded set $\hat{\mathcal{B}} \subset \mathcal{X}$.

Proof. Let $z \in \omega(\mathbb{R} \times \hat{\mathcal{B}})$. By definition, there exist sequences $\{t_k\}$, $\{(\hat{\tau}_k, \hat{x}_k)\}$, $k \in \mathbb{N}$ such that $t_k \to \infty$ as $k \to \infty$, $(\hat{\tau}_k, \hat{x}_k) \in \mathbb{R} \times \hat{\mathcal{B}}$, and $x(\hat{\tau}_k + t_k; (\hat{\tau}_k, \hat{x}_k)) \to z$ as $k \to \infty$, where $x(\hat{\tau}_k + t_k; (\hat{\tau}_k, \hat{x}_k))$ denotes the solution satisfying $x(\hat{\tau}_k) = \hat{x}_k$. Since all $\hat{x}_k \in \operatorname{Cl} \hat{\mathcal{B}}$, which is a compact subset of $\mathcal{X}$, there exists a time $T \in \mathbb{R}_+$ such that $x_k \triangleq x(\hat{\tau}_k + T; (\hat{\tau}_k, \hat{x}_k)) \in \mathcal{B}$ for all $k \in \mathbb{N}_N$. Define $\tau_k = \hat{\tau}_k - T$ and consider the sequence $\{(\tau_k, x_k)\}$ with $(\tau_k, x_k) \in \mathbb{R} \times \mathcal{B}$ for all $k \in \mathbb{N}_N$. We have $x(\hat{\tau}_k + t_k; (\hat{\tau}_k, \hat{x}_k)) = x(\tau_k + t_k, (\tau_k, x_k))$, i.e., $\lim_{k\to\infty} x(\tau_k + t_k; (\tau_k, x_k)) = z$, which shows that $z \in \omega(\mathbb{R} \times \mathcal{B})$. □

An immediate consequence of the above lemma is the following corollary:

Corollary 5.14. *Given a system $\dot{x}(t) = f(t, x(t))$ where $f : \mathbb{R} \times \mathbb{R}^n \to \mathbb{R}^n$ is piecewise continuous in the first argument and locally Lipschitz the second argument. Suppose $\mathcal{X} \subset \mathbb{R}^n$ is a closed set such that solutions $x(t_0 + t)$ with initial data $(t_0, x(t_0)) \subset \mathbb{R} \times \mathcal{X}$ exist and remain in $\mathcal{X}$ for all $t \in \overline{\mathbb{R}}_+$.*

Assume there are two bounded sets $\mathcal{B}_1 \subset \mathcal{X}$ and $\mathcal{B}_2 \subset \mathcal{X}$ with the property that for any compact subset $\mathcal{X}_0 \subset \mathcal{X}$ there exist times $T_1, T_2 \in \mathbb{R}_+$ such that solutions $x(t_0 + t)$ with initial data $(t_0, x(t_0)) \in \mathbb{R} \times \mathcal{X}_0$ satisfy $x(t_0 + t) \in \mathcal{B}_1$ for all $t \geq T_1$ and $x(t_0 + t) \in \mathcal{B}_2$ for all $t \geq T_2$.

Then $\omega(\mathbb{R} \times \mathcal{B}_1) = \omega(\mathbb{R} \times \mathcal{B}_2)$.

Proof. By Lemma 5.13, we have $\omega(\mathbb{R} \times \mathcal{B}_1) \subset \omega(\mathbb{R} \times \mathcal{B}_2)$ and $\omega(\mathbb{R} \times \mathcal{B}_1) \supset \omega(\mathbb{R} \times \mathcal{B}_2)$. □

Together, Lemma 5.13 and Corollary 5.14 show that the set $\omega(\mathbb{R} \times \mathcal{B}^*)$, where $\mathcal{B}^*$ is an ultimate bound for solutions of (5.14) with initial data $(t_0, x^*(t_0)) \in \mathbb{R} \times \mathcal{X}^*$, is uniquely defined despite $\mathcal{B}^*$ not being unique. Furthermore, the limit set of any bounded set contained in $\mathcal{X}^*$ is contained in $\omega(\mathbb{R} \times \mathcal{B}^*)$, i.e., $\omega(\mathbb{R} \times \mathcal{B}^*)$ contains all relevant limit sets. Thus $\omega(\mathbb{R} \times \mathcal{B}^*)$ is a candidate set for characterizing the steady state locus of a given system.

We aim at conditions that imply that the steady state outputs are synchronous. Therefore, similarly to the linear case, we define the synchronous system output set and the synchronous virtual output set

$$\mathcal{Y}_s \triangleq \left\{ y = (y_1^T, \ldots, y_N^T)^T \in \mathbb{R}^{Nq} \,|\, y_1 = \cdots = y_N \in \mathbb{R}^q \right\} \subset \mathbb{R}^{Nq},$$
$$\mathcal{W}_s \triangleq \left\{ \zeta = (\zeta_1^T, \ldots, \zeta_N^T)^T \in \mathbb{R}^{Nr} \,|\, \zeta_1 = \cdots = \zeta_N \in \mathbb{R}^r \right\} \subset \mathbb{R}^{Nr},$$

where all system outputs and all virtual outputs respectively are identical. Furthermore, we define the synchronous set

$$\mathcal{X}_s^* \triangleq \{ x^* \in \mathbb{R}^\sigma \,|\, h^*(x^*) \in \mathcal{Y}_s \wedge \beta^*(x^*) \in \mathcal{W}_s \} \subset \mathbb{R}^\sigma,$$

i.e., the set of states that are mapped to $\mathcal{Y}_s$ by $h^*(\cdot)$ and to $\mathcal{W}_s$ by $\beta^*(\cdot)$. Note that $\mathcal{X}_s^* \subset \mathbb{R}^\sigma$ is closed by continuity of the maps h^* and β^*. With these definitions, we are ready to state the first part of the result.

Theorem 5.15. *Let $\mathcal{G}(t) = \{\mathcal{V}, \mathcal{E}(t), W(t)\}$ be some communication graph with $|\mathcal{V}| = N > 1$. Let $f_k : \mathbb{R}^{n_k} \times \mathbb{R}^{p_k} \to \mathbb{R}^{n_k}$ and $h_k : \mathbb{R}^{n_k} \to \mathbb{R}^q$ be locally Lipschitz maps for $k \in \mathbb{N}_N$, $\mathcal{X}_k \subset \mathbb{R}^{n_k}$ be closed sets, and let $d : \mathbb{R}^r \times \mathbb{R}^r \to \mathbb{R}^r$ be a locally Lipschitz map satisfying $d(\zeta, \zeta) = 0$ for all $\zeta \in \mathbb{R}^r$*

Suppose the sets $\mathcal{Z}_k \subset \mathbb{R}^{m_k}$ and the maps $\phi_k : \mathbb{R}^{m_k} \times \mathbb{R}^r \times \mathbb{R}^q \to \mathbb{R}^{m_k}$, $\alpha_k : \mathbb{R}^{m_k} \times \mathbb{R}^r \times \mathbb{R}^q \to \mathbb{R}^{p_k}$, and $\beta_k : \mathbb{R}^{m_k} \times \mathbb{R}^q \to \mathbb{R}^r$ for $k \in \mathbb{N}_N$ are such that Properties (P1) and (P2) of the Nonlinear Heterogeneous Output Synchronization Problem 5.12 are satisfied.

Then Property (P3) is also satisfied if and only if $\omega(\mathbb{R} \times \mathcal{B}^*) \subset \mathcal{X}_s^*$.

Proof. ($\Leftarrow$) Let $\mathcal{X}_0^* \subset \mathcal{X}^*$ be any compact set. By Property (P2), there exists $T \in \mathbb{R}_+$ such that solutions $x(t_0+t)$ of (5.14) with initial data $(t_0, x(t_0)) \in \mathbb{R} \times \mathcal{X}_0^*$ satisfy $x(t_0+t) \in \mathcal{B}^*$ for all $t \geq T$. Thus $\omega(\mathbb{R} \times \mathcal{X}_0^*) \subset \omega(\mathbb{R} \times \mathcal{B}^*)$ by Lemma 5.13. Therefore $\omega(\mathbb{R} \times \mathcal{B}^*) \subset \mathcal{X}_s^*$ implies $\omega(\mathbb{R} \times \mathcal{X}_0^*) \subset \mathcal{X}_s^*$ and thus, by Lemma B.6 Property (P3).

($\Rightarrow$) Since $\mathcal{X}_0^*$ is arbitrary in Property (P3), it needs in particular to be satisfied for $\mathcal{X}_0^* = \operatorname{Cl} \mathcal{B}^*$. In that case $\omega(\mathbb{R} \times \mathcal{X}_0^*) = \omega(\mathbb{R} \times \mathcal{B}^*)$ by definition of limit sets. That is, $\omega(\mathbb{R} \times \mathcal{B}^*) \subset \mathcal{X}_s^*$ is also sufficient for Property (P3) by Lemma B.6. □

Theorem 5.15 states that any solution to Problem 5.12 is such that the set $\omega(\mathbb{R} \times \mathcal{B}^*)$ is contained in the set where the outputs and virtual outputs are identically synchronous. If we take Properties (P1) and (P2) as assumptions, this inclusion relation is even necessary and sufficient for given dynamic coupling controllers (5.12) to solve the problem. Hence, we know that there exists some attractive set possessing certain properties with respect to the system outputs $y(t)$ and the virtual outputs $\zeta(t)$. Yet, in order for the set to be a meaningful characterization of the steady state locus of the global system (5.14), we need some invariance properties, which are the subject of the next theorem.

Theorem 5.16. *Let $\mathcal{G}(t) = \{\mathcal{V}, \mathcal{E}(t), W(t)\}$ be some communication graph with $|\mathcal{V}| = N > 1$. Let $f_k : \mathbb{R}^{n_k} \times \mathbb{R}^{p_k} \to \mathbb{R}^{n_k}$ and $h_k : \mathbb{R}^{n_k} \to \mathbb{R}^q$ be locally Lipschitz maps for $k \in \mathbb{N}_N$, $\mathcal{X}_k \subset \mathbb{R}^{n_k}$ be closed sets, and let $d : \mathbb{R}^r \times \mathbb{R}^r \to \mathbb{R}^r$ be a locally Lipschitz map satisfying $d(\zeta, \zeta) = 0$ for all $\zeta \in \mathbb{R}^r$.*

Suppose the sets $\mathcal{Z}_k \subset \mathbb{R}^{m_k}$ and the maps $\phi_k : \mathbb{R}^{m_k} \times \mathbb{R}^r \times \mathbb{R}^q \to \mathbb{R}^{m_k}$, $\alpha_k : \mathbb{R}^{m_k} \times \mathbb{R}^r \times \mathbb{R}^q \to \mathbb{R}^{p_k}$, and $\beta_k : \mathbb{R}^{m_k} \times \mathbb{R}^q \to \mathbb{R}^r$ for $k \in \mathbb{N}_N$ solve the Nonlinear Heterogeneous Output Synchronization Problem 5.12.

Then $\omega(\mathbb{R}\times\mathcal{B}^*)$ *is a nonempty compact subset of* $\mathcal{X}^*$ *invariant for* (5.14a)–(5.14c) *with* $\delta(t)\equiv 0$.

Proof. We show that (5.14) is asymptotically invariant with limit system (5.14a)–(5.14c) with $\delta(t)\equiv 0$. Since $d(\cdot,\cdot)$ is continuous and $w_{k,j}(\cdot)$ is uniformly bounded by the definition of a communication graph, we know as a consequence of Property (P3) that $\delta(t)=(\delta_1(t)^T,\dots,\delta_N(t)^T)^T=D(t,\beta^*(x^*(t)))\to 0$ as $t\to\infty$ uniformly in initial data $(t_0,x^*(t_0))\in\mathbb{R}\times\mathcal{X}_0^*$ for any compact subset $\mathcal{X}_0^*\subset\mathcal{X}^*$. Thus, using the fact that $f_k^*(\cdot,\cdot)$ is locally Lipschitz, $f_k^*(x_k^*,\delta_k(t))\to f_k^*(x^*,0)$ as $t\to\infty$ uniformly in $x_k^*\in\mathcal{X}_{k,0}^*$ for all compact subsets $\mathcal{X}_{k,0}^*\subset\mathbb{R}^{n_k+m_k}$, and, by definition of $f^*(\cdot,\cdot)$, we thus have $f^*(x^*,\delta(t))\to f^*(x^*,0)$ as $t\to\infty$ uniformly in $x^*\in\mathcal{X}_0^*$ for any compact subset $\mathcal{X}_0^*\subset\mathbb{R}^\sigma$. The theorem then follows from Lemma B.7. □

Together with Theorem 5.15, we know by Theorem 5.16 that any solution to Problem 5.12 is such that the set $\omega(\mathbb{R}\times\mathcal{B}^*)$ is contained in the set where the outputs and virtual outputs are identically synchronous, it attracts all solutions of (5.14) uniformly for compact sets of initial conditions, and it is invariant for the uncoupled systems (5.14a)–(5.14c) with $\delta(t)\equiv 0$ (and thus also for the global system (5.14)). Furthermore, by Lemma 5.14, we know that the set $\omega(\mathbb{R}\times\mathcal{B}^*)$ is well defined in the sense that it does not depend on the choice of the set $\mathcal{B}^*$, which can be any bounded set that ultimately contains all solutions of (5.14) as described in Property (P2) of the problem statement. Note furthermore that $\omega(\mathbb{R}\times\mathcal{B}^*)$ is the smallest set with the above properties since any closed set that uniformly attracts solutions of (5.14) must contain $\omega(\mathbb{R}\times\mathcal{B}^*)$ by Lemma B.6, which we interpret as the property that solutions contained in $\omega(\mathbb{R}\times\mathcal{B}^*)$ persist. Thus, the set $\omega(\mathbb{R}\times\mathcal{B}^*)$ plays the role of the persistent invariant synchronous subspace $\mathcal{X}_p^*$ in the linear case and it is well suited to characterize the steady state locus of (5.14) (see Isidori and Byrnes, 2008, and the references given there).

Having settled the steady state *locus* of the network (5.14), we will now investigate the steady state *behavior*, i.e., the type of synchronous trajectories in the network. Motivated by the results obtained for the linear case, we assume that the synchronous outputs of the network (5.14) are generated by some virtual exosystem

$$\dot{\xi}(t)=s(\xi(t)) \tag{5.15a}$$
$$\eta(t)=\hat{h}(\xi(t)) \tag{5.15b}$$

with state vector $\xi(t)\in\hat{\mathcal{X}}\subset\mathbb{R}^\nu$ and output vector $\eta(t)\in\mathbb{R}^q$, where the subset $\hat{\mathcal{X}}\subset\mathbb{R}^\nu$ is compact and invariant. We say that the exosystem (5.15) *uniformly generates synchronous trajectories* of (5.14) if it possesses the property that (i) for any solution $x^*(t_0+t)$ to (5.14), there exists a solution $\xi(t)$ with initial data $\xi(0)\in\hat{\mathcal{X}}$ to the virtual exosystem such that $h^*(x^*(t_0+t))-1_N\otimes\hat{h}(\xi(t))\to 0$ as $t\to\infty$ uniformly in initial data $(t_0,x(t_0))\in\mathbb{R}\times\mathcal{X}_0^*$ for any compact subset $\mathcal{X}_0^*\subset\mathcal{X}^*$; and (ii) there exists a compact subset $\mathcal{X}_0^*\subset\mathcal{X}^*$ with the property that for any solution $\xi(t)$ to the virtual exosystem there exists a solution $x^*(t_0+t)$ to (5.14) with initial data $(t_0,x(t_0))\in\mathbb{R}\times\mathcal{X}_0^*$ such that $h^*(x^*(t_0+t))-1_N\otimes\hat{h}(\xi(t))\to 0$ as $t\to\infty$ uniformly in initial data $\xi(0)\in\hat{\mathcal{X}}$. These properties can be described formally as follows: (i) given any compact set $\mathcal{X}_0^*\subset\mathcal{X}^*$ and any $\varepsilon\in\mathbb{R}_+$, there exists $T\in\mathbb{R}_+$ such that for all initial conditions $(t_0,x^*(t_0))\in\mathbb{R}\times\mathcal{X}_0^*$ there exist initial conditions $\xi(0)\in\hat{\mathcal{X}}$ such that $\|h^*(x^*(t_0+t))-1_N\otimes\hat{h}(\xi(t))\|\le\varepsilon$ for all $t\ge T$ and (ii) there exists a compact subset $\mathcal{X}_0^* \subset \mathcal{X}^*$ with the property that for any $\varepsilon \in \mathbb{R}_+$, there exists $T \in \mathbb{R}_+$ such that for all initial conditions $\xi(0) \in \hat{\mathcal{X}}$ there exists initial conditions $(t_0, x^*(t_0)) \in \mathbb{R} \times \mathcal{X}_0^*$ such that $\|h^*(x^*(t_0+t)) - 1_N \otimes \hat{h}(\xi(t))\| \leq \varepsilon$ for all $t \geq T$. The first property says that any synchronous output of the network is generated by the virtual exosystem while the second property says that the virtual exosystem is minimal in the sense that it does not produce outputs that cannot be synchronous outputs for the network. In any case, we require convergence to be uniform for compact sets of initial conditions. The following result relates the exosystem (5.15) to the steady state behavior of the network (5.14):

Lemma 5.17. *Let $\mathcal{G}(t) = \{\mathcal{V}, \mathcal{E}(t), W(t)\}$ be some communication graph with $|\mathcal{V}| = N > 1$. Let $f_k : \mathbb{R}^{n_k} \times \mathbb{R}^{p_k} \to \mathbb{R}^{n_k}$ and $h_k : \mathbb{R}^{n_k} \to \mathbb{R}^q$ be locally Lipschitz maps for $k \in \mathbb{N}_N$, $\mathcal{X}_k \subset \mathbb{R}^{n_k}$ be closed sets, and let $d : \mathbb{R}^r \times \mathbb{R}^r \to \mathbb{R}^r$ be a locally Lipschitz map satisfying $d(\zeta, \zeta) = 0$ for all $\zeta \in \mathbb{R}^r$.*

Suppose the maps $\phi_k : \mathbb{R}^{m_k} \times \mathbb{R}^r \times \mathbb{R}^q \to \mathbb{R}^{m_k}$, $\alpha_k : \mathbb{R}^{m_k} \times \mathbb{R}^r \times \mathbb{R}^q \to \mathbb{R}^{p_k}$, $\beta_k : \mathbb{R}^{m_k} \times \mathbb{R}^q \to \mathbb{R}^r$ for $k \in \mathbb{N}_N$, and the sets $\mathcal{Z}_k \subset \mathbb{R}^{m_k}$ for $k \in \mathbb{N}_N$ solve the Nonlinear Heterogeneous Output Synchronization Problem 5.12.

Suppose the compact set $\hat{\mathcal{X}} \subset \mathbb{R}^\nu$ and the maps $s : \hat{\mathcal{X}} \to \hat{\mathcal{X}}$ and $\hat{h} : \hat{\mathcal{X}} \to \mathbb{R}^q$ are such that $\hat{\mathcal{X}}$ is invariant for (5.15a), *and* (5.15) *uniformly generates synchronous trajectories for* (5.14).

Then for any solution $x^*(t)$ to (5.14) *contained in $\omega(\mathbb{R} \times \mathcal{B}^*)$, there exists a solution $\xi(t)$ to* (5.15) *contained in $\hat{\mathcal{X}}$ and conversely for any solution $\xi(t)$ to* (5.15) *contained in $\hat{\mathcal{X}}$, there exists a solution $x^*(t)$ to* (5.14) *contained in $\omega(\mathbb{R} \times \mathcal{B}^*)$ such that $h^*(x^*(t)) = 1_N \otimes \hat{h}(\xi(t))$ along these solutions.*

Proof. Denote the solution of (5.14) satisfying $x^*(t_0) = x_0^*$ as $x^*(t; (t_0, x_0))$ and the solution of (5.15) satisfying $\xi(0) = \xi_0$ as $\xi(t; \xi_0)$. Suppose (5.15) uniformly generates synchronous outputs for (5.14).

Then, given any compact set $\mathcal{X}_0^* \subset \mathcal{X}^*$, there exists a map $c_1 : \mathbb{R} \times \mathcal{X}_0^* \to \hat{\mathcal{X}}$ such that $\left[h^*(x^*(t_0+t; (t_0, x_0^*))) - \hat{h}^*(\xi(t; c_1(t_0, x^*(t_0))))\right] \to 0$ as $t \to \infty$ uniformly in $(t_0, x_0^*) \in \mathbb{R} \times \mathcal{X}_0^*$. That is, solutions to the composite system (5.14), (5.15) with initial data in the set $\mathcal{T}_1 \triangleq \{(t_0, x_0^*, \xi_0) \in \mathbb{R} \times \mathcal{X}_0^* \times \hat{\mathcal{X}} | \xi_0 = c_1(t_0, x_0^*)\}$ uniformly converge to the set $\tilde{\mathcal{X}} \triangleq \{(x^*, \xi) \in \mathcal{X}^* \times \hat{\mathcal{X}} | h^*(x^*) = \hat{h}(\xi)\}$. By Lemma B.6, this implies that $\omega(\mathcal{T}_1) \subset \tilde{\mathcal{X}}$. Choose $\mathcal{X}_0^* = \omega(\mathbb{R} \times \mathcal{B}^*)$. Since the two subsystems (5.14), (5.15) are decoupled, $\omega(\mathcal{T}_1) = \omega(\mathbb{R} \times \mathcal{B}^*) \times \omega(c_1(\mathbb{R}, \omega(\mathbb{R} \times \mathcal{B}^*))) \subset \omega(\mathbb{R} \times \mathcal{B}^*) \times \hat{\mathcal{X}}$. The forward direction of the statement follows.

On the other hand, there exists a set $\mathcal{X}_0^*$ and a map $c_2 : \hat{\mathcal{X}} \to \mathbb{R} \times \mathcal{X}_0^*$ such that $\left[h^*(x^*(t_0+t; c_2(\xi_0))) - \hat{h}^*(\xi(t; \xi_0))\right] \to 0$ as $t \to \infty$ uniformly in $(t_0, x_0^*) \in \mathbb{R} \times \mathcal{X}_0^*$. That is, solutions to the composite system (5.14), (5.15) with initial data in the set $\mathcal{T}_2 \triangleq \{(t_0, x_0^*, \xi_0) \in \mathbb{R} \times \mathcal{X}_0^* \times \hat{\mathcal{X}} | (t_0, x_0^*) = c_2(\xi_0)\}$ uniformly converge to the set $\tilde{\mathcal{X}}$. By Lemma B.6, this implies that $\omega(\mathcal{T}_2) \subset \tilde{\mathcal{X}}$. Since the two subsystems (5.14), (5.15) are decoupled, $\omega(\mathcal{T}_2) = \omega(c_2(\hat{\mathcal{X}})) \times \hat{\mathcal{X}} \subset \omega(\mathbb{R} \times \mathcal{B}^*) \times \hat{\mathcal{X}}$. The backwards direction of the statement follows. □

By Lemma 5.17, we know that any virtual exosystem, which uniformly generates synchronous output trajectories for the network (5.14), completely characterizes the steady state output behavior of (5.14). A candidate for a virtual exosystem (5.15) that uniformly generates synchronous outputs is given by $s(\xi(t)) = f^*(\xi(t), 0)$, $\hat{h}(\xi(t)) = \Upsilon_k h^*(\xi(t))$, and

$\hat{\mathcal{X}} = \omega(\mathbb{R} \times \mathcal{B}^*)$, where $\Upsilon_k : \mathbb{R}^{Nq} \to \mathbb{R}^q$ is the projection defined as $y = (y_1^T, \ldots, y_N^T)^T \mapsto y_k$ for $k \in \mathbb{N}_N$. Since $\omega(\mathbb{R} \times \mathcal{B}^*) \subset \mathcal{X}_s^*$, the choice of $k \in \mathbb{N}_N$ is arbitrary. However, even though this virtual exosystem is simply a copy of the global system in steady state, it is not guaranteed to uniformly generate synchronous outputs. This is because uniform convergence to some set does not imply convergence to a particular solution contained in that set in general. Hence, we will formulate existence of a virtual exosystem that uniformly generates synchronous outputs as an assumption in the results below.

To state the final version of the implicit internal model principle, we need in addition some observability assumption as follows:

Assumption 5.18. *There exists some $\rho \in \mathbb{N} \bigcup \{0\}$ such that the map $f^*(\cdot, 0) : \mathbb{R}^\sigma \to \mathbb{R}^\sigma$ restricted to $\omega(\mathbb{R} \times \mathcal{B}^*)$ is $\rho - 1$ time continuously differentiable, the map $h^* : \mathbb{R}^\sigma \to \mathbb{R}^{Nq}$ restricted to $\omega(\mathbb{R} \times \mathcal{B}^*)$ is ρ times continuously differentiable, and there exists a map $\phi : \mathbb{R}^{\rho N q} \to \mathbb{R}^\sigma$ with the property that*

$$x^* = \phi\left(h^*(x^*), L_{f^*}h^*(x^*), \ldots, L_{f^*}^{\rho}h^*(x^*)\right)$$

for all $x^* \in \omega(\mathbb{R} \times \mathcal{B}^*)$.

Assumption 5.18 represents a strong observability requirement imposed on the network (5.14) in steady state. Namely, we require that in steady state, the state vector can be uniquely reconstructed from the outputs and their derivatives up to some order ρ and thus in particular that output derivatives exist up to that order. Since this requirement is restricted to the steady state, it can be interpreted as detectability of the network on the set $\mathcal{X}^*$ defined in Property (P1) of the problem statement. Assumption 5.18 is thus a nonlinear analogue for the detectability assumption in the linear case.

In the theorem below, we will need the projections $\Xi_k : \mathbb{R}^\sigma \to \mathbb{R}^{n_k+m_k}$ defined as $x^* = ((x_1^*)^T, \ldots, (x_N^*)^T)^T \mapsto x_k^*$ for all $k \in \mathbb{N}_N$ and the sets $\mathcal{X}_{k,p}^* \triangleq \Xi_k \omega(\mathbb{R} \times \mathcal{B}^*)$.

Theorem 5.19. *Let $\mathcal{G}(t) = \{\mathcal{V}, \mathcal{E}(t), W(t)\}$ be some communication graph with $|\mathcal{V}| = N > 1$. Let $f_k : \mathbb{R}^{n_k} \times \mathbb{R}^{p_k} \to \mathbb{R}^{n_k}$ and $h_k : \mathbb{R}^{n_k} \to \mathbb{R}^q$ be locally Lipschitz maps for $k \in \mathbb{N}_N$, $\mathcal{X}_k \subset \mathbb{R}^{n_k}$ be closed sets, and let $d : \mathbb{R}^r \times \mathbb{R}^r \to \mathbb{R}^r$ be a locally Lipschitz map satisfying $d(\zeta, \zeta) = 0$ for all $\zeta \in \mathbb{R}^r$.*

Suppose the maps $\phi_k : \mathbb{R}^{m_k} \times \mathbb{R}^r \times \mathbb{R}^q \to \mathbb{R}^{m_k}$, $\alpha_k : \mathbb{R}^{m_k} \times \mathbb{R}^r \times \mathbb{R}^q \to \mathbb{R}^{p_k}$, $\beta_k : \mathbb{R}^{m_k} \times \mathbb{R}^q \to \mathbb{R}^r$ for $k \in \mathbb{N}_N$, and the sets $\mathcal{Z}_k \subset \mathbb{R}^{m_k}$ for $k \in \mathbb{N}_N$ solve the Nonlinear Heterogeneous Output Synchronization Problem 5.12 and Assumption 5.18 holds.

Suppose furthermore that there exists a compact set $\hat{\mathcal{X}} \subset \mathbb{R}^\nu$ and maps $s : \hat{\mathcal{X}} \to \hat{\mathcal{X}}$ and $\hat{h} : \hat{\mathcal{X}} \to \mathbb{R}^q$ such that $\hat{\mathcal{X}}$ is invariant for (5.15a) *and the virtual exosystem* (5.15) *uniformly generates synchronous outputs for the network* (5.14).

Then there exist surjective maps $\psi_k : \hat{\mathcal{X}} \to \mathcal{X}_{k,p}^*$, $k \in \mathbb{N}_N$ with the property that the sets $\{(\xi, x_k^*) \in \hat{\mathcal{X}} \times \mathcal{X}_{k,p}^* | x_k^* = \psi_k(\xi)\}$, $k \in \mathbb{N}_N$ are invariant sets for (5.15), (5.13) *with $\delta_k(t) \equiv 0$ and*

$$\frac{\partial \psi_k(\xi)}{\partial \xi} s(\xi) = f_k^*(\psi_k(\xi), 0) \tag{5.16a}$$

$$\hat{h}(\xi) = h_k^*(\psi_k(\xi)) \tag{5.16b}$$

for all $\xi \in \hat{\mathcal{X}}$ in case of (5.16b) *and for all $\xi \in \hat{\mathcal{X}}$ at which $\psi_k(\xi)$ is continuously differentiable in case of* (5.16a) *for all $k \in \mathbb{N}_N$.*

Proof. By virtue of Lemma 5.17, we know that for any solution $x^*(t)$ to (5.14) contained in $\omega(\mathbb{R}\times\mathcal{B}^*)$, there exists a solution $\xi(t)$ to (5.15) contained in $\hat{\mathcal{X}}$ with the property that $y(t) = 1_N \otimes \eta(t)$ along those solutions for all $t \in \mathbb{R}$. Furthermore, by Assumption 5.18, we know that there exists some $\rho \in \mathbb{N}\bigcup\{0\}$ such that the derivatives $y^{(j)}(t)$ exist along the solution $x^*(t)$ for all $j \in \mathbb{N}_N\bigcup\{0\}$. Thus, the derivatives $\eta^{(j)}(t)$ along the solution $\xi(t)$ exist as well and $y^{(j)}(t) = 1_N\otimes\eta^{(j)}(t)$ for all $j \in \mathbb{N}_N\bigcup\{0\}$. This shows that the derivatives $y^{(j)}(t)$ are uniquely determined as functions of the exosystem state $\xi(t)$. However, by Assumption 5.18 those derivatives uniquely determine $x^*(t)$, which in turn implies that there exists a map $\psi : \hat{\mathcal{X}} \to \omega(\mathbb{R}\times\mathcal{B}^*)$ such that along the solutions $x^*(t)$ and $\xi(t)$, $x^*(t) = \psi(\xi(t))$ holds identically. Since the solution $x^*(t)$ is any solution in $\omega(\mathbb{R}\times\mathcal{B}^*)$ and the equality holds for all times $t \in \mathbb{R}$, the map $\psi(\cdot)$ is surjective and the set $\{(x^*,\xi) \in \omega(\mathbb{R}\times\mathcal{B}^*)\times\hat{\mathcal{X}}|x^* = \psi(\xi)\}$ is invariant for (5.15), (5.14) with $\delta(t) \equiv 0$. Furthermore, $1_N \otimes \hat{h}(\xi) = h^*(\psi(\xi))$ for all $\xi \in \hat{\mathcal{X}}$ by construction of $\psi(\cdot)$.

Define the maps $\psi_k : \hat{\mathcal{X}} \to \mathcal{X}^*_{k,p}$, $k \in \mathbb{N}_N$ as $\psi_k(\xi) \triangleq \Xi_k\psi(\xi)$, where Ξ_k is the projection from $\mathbb{R}^\sigma$ onto $\mathbb{R}^{n_k\times m_k}$ defined before. The maps $\psi_k(\cdot)$ thus defined are surjective. And, since the individual systems (5.13) are decoupled if $x^* \in \omega(\mathbb{R}\times\mathcal{B}^*)$, the sets $\{(x_k^*,\xi) \in \mathcal{X}^*_{k,p}\times\hat{\mathcal{X}}|x_k^* = \psi_k(\xi)\}$, $k \in \mathbb{N}_N$ need to be invariant for (5.15), (5.13) with $\delta_k(t) \equiv 0$. Condition (5.16a) is simply an infinitesimal characterization of this invariance obtained from the observation that $x_k^*(t) = \psi_k(\xi(t))$ for all $t \in \mathbb{R}$ implies $\dot{x}_k^*(t) = \frac{\mathrm{d}}{\mathrm{d}t}\psi_k(\xi(t))$, if the derivatives exist. To see that (5.16b) also holds, it is enough to note that $h^*(x^*) = \left((h_1^*(\Xi_1x^*))^T,\ldots,(h_N^*(\Xi_Nx^*))^T\right)^T$ by definition of $h^*(\cdot)$. This completes the proof. □

This final version of the implicit internal model principle conveys a rather strong statement. Namely, whenever the Nonlinear Heterogeneous Output Synchronization Problem (5.14) is solved and the synchronous outputs are uniformly generated by some virtual exosystem (5.15), each of the individual systems with its dynamic coupling controller (5.13) must be able to produce all steady state outputs of the exosystem, i.e., it needs to have an internal model of the virtual exosystem embedded in its dynamics. Conditions (5.16) are a mathematical statement of this fact. Assumption 5.18 is needed in the theorem to make the maps $\psi_k(\cdot)$, $k \in \mathbb{N}_N$ unique by imposing that the states x_k^* are uniquely determined by outputs and output derivatives. The result is then a consequence of the fact that the exosystem states uniquely determine the exosystem outputs and their derivatives and the property that exosystem solutions produce the same output trajectories as the synchronized network.

The result is illustrated in Figure 5.4 for the case when the network synchronizes to some oscillatory trajectories. In that case, it is possible to define the virtual exosystem on the circle $\hat{\mathcal{X}} = \mathbb{S}^1$, and $\psi_k(\cdot)$ maps the circle to the limit cycles of the individual systems. We pursue the example of the oscillators a little further below.

Example 5.20. *Consider a network of N systems with static diffusive couplings, i.e., $m_k = 0$, $x_k^* = x_k$, and $f_k^*(x_k^*,\delta_k) = f_k(x_k,\alpha_k(\delta_k))$, and assume the systems synchronize and the synchronous outputs are uniformly generated by the virtual exosystem given as*

$$\dot{\xi}(t) = \begin{pmatrix} 0 & -\omega_0 \\ \omega_0 & 0 \end{pmatrix}\xi(t)$$
$$\eta(t) = \arg(\xi_1(t) + \mathrm{j}\xi_2(t))$$

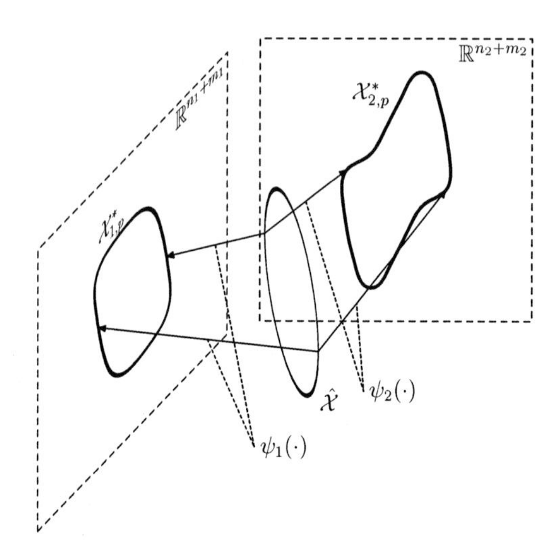

Figure 5.4: State spaces $\mathbb{R}^{n_1+m_1}$, $\mathbb{R}^{n_2+m_2}$ of two individual systems with dynamic coupling controllers with invariant sets $\mathcal{X}^*_{1,p}$ and $\mathcal{X}^*_{2,p}$ and steady state set $\hat{\mathcal{X}}$ of the virtual exosystem.

with $\hat{\mathcal{X}} = \{\xi \in \mathbb{R}^2 | \|\xi\| = 1\}$. *Then we know by Theorem 5.19 that the individual systems admit compact invariant subspaces* $\mathcal{X}^*_{k,p} \subset \mathbb{R}^{n_k+m_k}$ *that are mapped to the circle by* $h^*_k(\cdot)$, *i.e.,* $h^*_k(\mathcal{X}^*_{k,p}) = \mathbb{S}^1$, *with the property that* $L_{f^*_k} h^*_k(x^*_k) = \omega$ *for all* $x^*_k \in \mathcal{X}^*_{k,p}$.

In particular, we know that a group of N *oscillators can be synchronized by means of static diffusive couplings only if they all have the same frequency of oscillation.*

The results presented so far are the nonlinear analogue to the implicit internal model principle derived in the linear case in Section 5.1.2. In fact, it is easily verified that conditions derived in Theorem 5.19 exactly reduce to the conditions given in Theorem 5.4 if the network is linear. As a next step, in parallel to the presentation of the linear case in Section 5.1.2, we derive explicit conditions for solvability of the Nonlinear Heterogeneous Output Synchronization Problem 5.12.

Explicit Internal Model Principle

We have the following Lemma, analogue to Lemma 5.7 for the linear case:

Lemma 5.21. *Let* $k \in \mathbb{N}_N$ *fixed. Let* $f_k : \mathbb{R}^{n_k} \times \mathbb{R}^{p_k} \to \mathbb{R}^{n_k}$, $h_k : \mathbb{R}^{n_k} \to \mathbb{R}^q$, $s : \hat{\mathcal{X}} \to \hat{\mathcal{X}}$, *and* $\hat{h} : \hat{\mathcal{X}} \to \mathbb{R}^q$, *where* $\hat{\mathcal{X}} \subset \mathbb{R}^\nu$ *is compact and invariant for* (5.15a).

There exist maps $\phi_k : \mathbb{R}^{m_k} \times \mathbb{R}^r \times \mathbb{R}^q \to \mathbb{R}^{m_k}$, $\alpha_k : \mathbb{R}^{m_k} \times \mathbb{R}^r \times \mathbb{R}^q \to \mathbb{R}^{p_k}$, $\beta_k : \mathbb{R}^{m_k} \times \mathbb{R}^q \to \mathbb{R}^r$, *and* $\psi_k : \hat{\mathcal{X}} \to \mathcal{X}^*_{k,p}$ *such that* (5.16) *holds (with* $f^*_k : \mathbb{R}^{n_k+m_k} \times \mathbb{R}^r \to \mathbb{R}^{n_k+m_k}$ *and* $h^*_k : \mathbb{R}^{n_k+m_k} \to \mathbb{R}^q$ *as defined before) if and only if there exists a surjective map* $\pi_k : \hat{\mathcal{X}} \to \hat{\mathcal{X}}_k$ *with* $\hat{\mathcal{X}}_k = \{x_k \in \mathbb{R}^k \,|\, (x_k^T, z_k^T)^T \in \mathcal{X}^*_{k,p}, \, z_k \in \mathbb{R}^{m_k}\}$ *and a map* $\lambda_k : \hat{\mathcal{X}} \to \mathbb{R}^{p_k}$

such that

$$\frac{\partial \pi_k(\xi)}{\partial \xi} s(\xi) = f_k(\pi_k(\xi), \lambda_k(\xi)) \tag{5.17a}$$

$$\hat{h}(\xi) = h_k(\pi_k(\xi)) \tag{5.17b}$$

for all $\xi \in \hat{\mathcal{X}}$ *in case of* (5.17b) *and for all* $\xi \in \hat{\mathcal{X}}$ *at which* $\pi_k(\xi)$ *is continuously differentiable in case of* (5.17a).

Proof. Assume $\psi_k : \hat{\mathcal{X}} \to \mathcal{X}^*_{k,p}$ solves (5.16). Decompose $\psi_k(\cdot)$ as $\psi_k(\xi) = \left(\pi_k^T(\xi), \sigma_k^T(\xi)\right)^T$ with $\pi_k : \hat{\mathcal{X}} \to \hat{\mathcal{X}}_k$ and $\sigma_k : \hat{\mathcal{X}} \to \mathbb{R}^{m_k}$ and observe that $\pi_k(\cdot)$ is surjective. Substituting the definitions of f_k^* and h_k^* into (5.16), we obtain

$$\begin{aligned}
\frac{\partial \pi_k(\xi)}{\partial \xi} s(\xi) &= f_k\Big(\pi_k(\xi), \alpha_k\big(\sigma_k(\xi), 0, h_k(\pi_k(\xi))\big)\Big), \\
\frac{\partial \sigma_k(\xi)}{\partial \xi} s(\xi) &= \phi_k\big(\sigma_k(\xi), 0, h_k(\pi_k(\xi))\big), \\
\hat{h}(\xi) &= h_k(\pi_k(\xi)).
\end{aligned}$$

Defining $\lambda_k(\xi) \triangleq \alpha_k\big(\sigma_k(\xi), 0, h_k(\pi_k(\xi))\big) = \alpha_k\big(\sigma(\xi), 0, \hat{h}(\xi)\big)$ shows necessity. To prove sufficiency, we thus need to show that

$$\begin{aligned}
\frac{\partial \sigma_k(\xi)}{\partial \xi} s(\xi) &= \phi_k\big(\sigma(\xi), 0, \hat{h}(\xi)\big), \\
\lambda(\xi) &= \alpha_k\big(\sigma(\xi), 0, \hat{h}(\xi)\big)
\end{aligned}$$

admit a solution $\sigma_k : \hat{\mathcal{X}} \to \mathbb{R}^{m_k}$, $\phi_k(\cdot, 0, \cdot) : \mathbb{R}^{m_k} \times \mathbb{R}^q \to \mathbb{R}^{m_k}$, and $\lambda_k(\cdot, 0, \cdot) : \mathbb{R}^{m_k} \times \mathbb{R}^q \to \mathbb{R}^q$. One such solution is given as $\sigma_k(\xi) = \xi$, $\phi_k(\xi, 0, \eta) = s(\xi)$ and $\alpha_k(\xi, 0, \eta) = \lambda_k(\xi)$. Defining $\mathcal{X}^*_{k,p} = \left\{ (x_k^T, z_k^T)^T \in \mathbb{R}^{n_k+m_k} \,\middle|\, x_k = \pi_k(\xi), z_k = \sigma_k(\xi), \xi \in \hat{\mathcal{X}} \right\}$ and $\psi_k(\xi) = \left(\pi_k^T(\xi), \sigma_k^T(\xi)\right)^T$ completes the proof. □

As in the linear case, the dynamic couplings proposed in the proof of sufficiency represent purely feedforward control for the individual systems with the property that invariance of the steady state set, i.e., $\mathcal{X}^*_{k,p}$, is guaranteed. This feedforward coupling will usually be supplemented by some stabilizing feedback in order to solve the Nonlinear Heterogeneous Output Synchronization Problem 5.12.

With the help of the above result, we are finally able to state the explicit solvability conditions for the Nonlinear Heterogeneous Output Synchronization Problem 5.12.

Corollary 5.22. *Let* $\mathcal{G}(t) = \{\mathcal{V}, \mathcal{E}(t), W(t)\}$ *be some communication graph with* $|\mathcal{V}| = N > 1$. *Let* $f_k : \mathbb{R}^{n_k} \times \mathbb{R}^{p_k} \to \mathbb{R}^{n_k}$ *and* $h_k : \mathbb{R}^{n_k} \to \mathbb{R}^q$ *be locally Lipschitz maps for* $k \in \mathbb{N}_N$, $\mathcal{X}_k \subset \mathbb{R}^{n_k}$ *be closed sets, and let* $d : \mathbb{R}^r \times \mathbb{R}^r \to \mathbb{R}^r$ *be a locally Lipschitz map satisfying* $d(\zeta, \zeta) = 0$ *for all* $\zeta \in \mathbb{R}^r$.

Suppose the sets $\mathcal{Z}_k \subset \mathbb{R}^{m_k}$ *and the maps* $\phi_k : \mathbb{R}^{m_k} \times \mathbb{R}^r \times \mathbb{R}^q \to \mathbb{R}^{m_k}$, $\alpha_k : \mathbb{R}^{m_k} \times \mathbb{R}^r \times \mathbb{R}^q \to \mathbb{R}^{p_k}$, *and* $\beta_k : \mathbb{R}^{m_k} \times \mathbb{R}^q \to \mathbb{R}^r$ *for* $k \in \mathbb{N}_N$ *solve the Nonlinear Heterogeneous Output Synchronization Problem 5.12 and Assumption 5.18 holds.*

Suppose furthermore that there exists a compact set $\hat{\mathcal{X}} \subset \mathbb{R}^{\nu}$ *and maps* $s : \hat{\mathcal{X}} \to \hat{\mathcal{X}}$ *and* $\hat{h} : \hat{\mathcal{X}} \to \mathbb{R}^{q}$ *such that* $\hat{\mathcal{X}}$ *is invariant for* (5.15a) *and the virtual exosystem* (5.15) *uniformly generates synchronous outputs for the network* (5.14).

Then there exist surjective maps $\pi_k : \hat{\mathcal{X}} \to \hat{\mathcal{X}}_k$ *and maps* $\lambda_k : \hat{\mathcal{X}} \to \mathbb{R}^{p_k}$*, for all* $k \in \mathbb{N}_N$*, with* $\hat{\mathcal{X}}_k$ *as defined in Lemma 5.21, that solve* (5.17b) *for all* $\xi \in \hat{\mathcal{X}}$ *and* (5.17a) *for all* $\xi \in \hat{\mathcal{X}}$ *for which* $\pi_k(\xi)$ *is continuously differentiable.*

Proof. By Lemma 5.21, solvability of (5.16) implies solvability of (5.17). The corollary thus follows from Theorem 5.19. □

The conditions (5.17) given in Lemma 5.21 are known as FBI equations (named after Bruce A. Francis, Christopher I. Byrnes, and Alberto Isidori) in the context of output regulation (see Byrnes and Isidori, 1998, 2003, Byrnes et al., 1997, Isidori and Byrnes, 2008, Knobloch et al., 1993, Krener, 1999, Pavlov et al., 2006). The systems theoretic interpretation of the FBI conditions is similar to the linear case. Namely, conditions (5.17) state that the individual systems (5.11) admit a controlled invariant set with the property that the restriction of the controlled system to that set is characterized by the virtual exosystem. While it was easy to solve the corresponding conditions in the linear case, for the nonlinear case, the conditions manifest in the form of partial differential equations which are hard to solve in general. However, even in cases when one does not find a solution to the FBI equations, one might still use the fact that they are guaranteed to exist under the assumptions and hypothesis given in the theorems.

There is one important difference in the nonlinear case as compared to the linear case which is worth being mentioned. In the linear case we showed that if the Heterogeneous Output Synchronization Problem 5.1 is solved, then the Francis equations admit a solution for a given virtual exosystem *if and only if* the virtual exosystem generates all synchronous outputs. In the nonlinear case we merely showed that if the Nonlinear Heterogeneous Output Synchronization Problem 5.12 is solved the FBI equations admit a solution for a given virtual exosystem *if* the virtual exosystem generates all synchronous outputs. In fact, the converse is not true in general. The reason for this is that in nonlinear systems convergence of a solution to a given invariant set does not imply convergence to a particular solution contained in this set. Therefore, for nonlinear systems, convergence to the steady state locus does not imply convergence to a steady state solution. To have this additional property, often referred to as the asymptotic phase property (see Coddington and Levinson, 1955, Hartman, 1964, Knobloch and Aulbach, 1981), additional assumptions are needed. Usually, one imposes (or simply assumes) that the convergence to the steady state locus is locally exponentially fast to enforce the asymptotic phase property. However, since we were interested in *necessary* conditions in terms of existence of an appropriate steady state locus in the preceding discussion, we ignored the subtlety of asymptotic phase properties of the steady state locus.

Finally, we would like to mention that Remark 5.9 given in the linear case is identically true for the nonlinear case. That is, solvability of the Nonlinear Heterogeneous Output Synchronization Problem 5.12 generally requires dynamic couplings.

5.3 Summary

In this chapter, we derived necessary conditions for synchronization in networks of individuals with non-identical dynamics. The first part of the chapter was dedicated to linear networks while the second part generalized the problem to nonlinear networks. The conditions were obtained as a consequence of one of the most fundamental paradigms for feedback control, which can be paraphrased as follows: If a certain property (in our case synchronization) shall be satisfied asymptotically, the system needs to possess a steady state locus, i.e., an invariant subset of the state space, which attracts all solutions and which is such that the system in steady state identically satisfies the property under consideration. Thus, whenever designing a feedback controller with the objective to asymptotically satisfy a given property, the first requirement, which the controller needs to comply with, is the existence of an appropriate steady state locus such that, in steady state, the property is identically satisfied. The second requirement is to make the invariant set representing the steady state attractive. The most simple application of this paradigm is stabilization of a single point, which is only possible if that point is an equilibrium for the closed loop system. More advanced problems include asymptotic tracking and disturbance rejection, summarized under the term *output regulation*, where a given system shall asymptotically track or reject signals generated by some exosystem. Again, for this to be possible, an appropriate steady state locus must exist for the closed loop system such that in steady state, the exosystem signals are identically tracked or rejected.

We showed in this chapter (see also Wieland and Allgöwer, 2009a,b, Wieland et al., 2010c) that the same kind of reasoning applies in the heterogeneous synchronization problem. If the network asymptotically synchronizes, the global system needs to possess an appropriate steady state locus such that the system in steady state exhibits identically synchronous outputs. The results in this chapter are essentially existence conditions for such a steady state locus and characterizations of the steady state behavior in terms of *internal model* interpretations. By imposing the constraint of exchanging only relative information over the network, i.e., the individual systems being decoupled once synchronized, the conditions apply to all individual systems with their local controllers independently. In fact, we derived conditions on the individual systems that correspond to the well-known Francis equations for the linear case and FBI equations in the nonlinear case, which show that a network can synchronize to the outputs of some dynamical system – called virtual exosystem in analogy with the exosystem appearing in output regulation – if the dynamic models of all individual systems together with their local coupling controllers embed an *internal model of the virtual exosystem.* The Francis equations and FBI equations give necessary conditions for a coupling controller to exist such that this internal model requirement is satisfied. They manifest in terms of linear matrix equations in the linear case and nonlinear partial differential equations in the nonlinear case. Since we were interested in synchronization over time-varying communication topologies, we derived all results for non-autonomous systems.

It has been argued that both in the linear case and in the nonlinear case, dynamic couplings are necessary to synchronize non-identical systems, because they are necessary to guarantee existence of an appropriate steady state locus. This observation supports the claim of this thesis, that dynamic couplings are necessary for solvability of consensus and synchronization problems with increased system *and* topological complexity.

Chapter 6

Synchronization in Heterogeneous Networks with Dynamic Couplings

In Chapter 5 we extensively discussed the fact that a network of individual systems with dynamic coupling controllers can achieve output synchronization by exchange of relative information, i.e., diffusive-like couplings, only if an internal model of some virtual exosystem is embedded in all individual systems together with their dynamic coupling controllers. It was shown that, in both the linear and the nonlinear case, dynamic couplings are generally necessary to enforce existence of appropriate steady states of the individual systems together with their dynamic coupling controllers. More precisely, we showed how the embedding of the internal model of a virtual exosystem in each individual system with its dynamic coupling controller yields an invariant set for the global system with the property that the dynamics restricted to this set exhibit identically synchronized outputs of the individual systems. The part which is missing to the solution of the heterogeneous output synchronization problems posed in the previous chapter, is the design of couplings that render this synchronous steady state set attractive. This is the subject of the present chapter.

To this end, the general approach will be to reduce the actual problem, i.e., making the synchronous steady state attractive, to simpler consensus and synchronization problems. Namely, in both the linear and the nonlinear case the problem will be reduced to a consensus problem with static diffusive couplings among linear systems, as it has been considered in Chapter 3. The solution to the heterogeneous output synchronization problem is then obtained by applying the methods presented there. In the following, we will consider strong communication constraints, i.e., we are interested in synchronization over uniformly connected communication graphs without any further assumptions on the communication topology. Hence, we are considering problems with great topological complexity.

We present solutions for two cases below: First, we consider solutions to the Linear Heterogeneous Output Synchronization Problem 5.1 in a very general way in Section 6.1. The results presented there have been published in Wieland et al. (2010c). It will turn out that under the assumption that the individual systems are all stabilizable and detectable and the synchronous outputs are polynomially bounded, the internal model conditions that were shown to be necessary for solvability of the linear heterogeneous output synchronization problem in Corollary 5.8 are also sufficient. Afterwards, in Section 6.2, we consider solutions to the Nonlinear Heterogeneous Output Synchronization Problem 5.12 in the particular case when the individual systems are exponentially stable oscillators.

The latter result is based on Wieland et al. (2010b), where the mechanism is illustrated for identical oscillators. We will show that, as long as the heterogeneity in the network is small enough (in a sense to be defined), solutions to the nonlinear heterogeneous output synchronization problem exist.

6.1 Synchronization in Networks of Non-Identical Linear Systems

6.1.1 Problem Statement

The problem setup is similar to the one presented in Section 5.1 and repeated here for the reader's convenience. As in Section 5.1, we consider a network of N different LTI systems modeled as

$$\dot{x}_k(t) = A_k x_k(t) + B_k u_k(t) \tag{6.1a}$$
$$y_k(t) = C_k x_k(t) \tag{6.1b}$$

with state vector $x_k(t) \in \mathbb{R}^{n_k}$, input vector $u_k(t) \in \mathbb{R}^{p_k}$ and output vector in $y_k(t) \in \mathbb{R}^q$ for $k \in \mathbb{N}_N$. The objective is to exponentially synchronize the outputs $y_k(t)$, $k \in \mathbb{N}_N$ by means of dynamic couplings

$$\dot{z}_k(t) = E_k z_k(t) + F_k \delta_k(t) + M_k y_k(t) \tag{6.2a}$$
$$u_k(t) = G_k z_k(t) + H_k \delta_k(t) + O_k y_k(t) \tag{6.2b}$$
$$\zeta_k(t) = P_k z_k(t) + Q y_k(t) \tag{6.2c}$$
$$\delta_k(t) = \sum_{j=1}^{N} w_{k,j}(t)(\zeta_k(t) - \zeta_j(t)) \tag{6.2d}$$

with coupling states $z_k(t) \in \mathbb{R}^{m_k}$ and virtual outputs $\zeta_k(t) \in \mathbb{R}^r$ for $k \in \mathbb{N}_N$. The controllers are driven by the system outputs $y_k(t)$ and the relative controller output signals $\delta_k(t)$. The outputs of the controller are the system input $u_k(t)$ and the virtual output $\zeta_k(t)$ (see also the block diagram given in Figure 5.1(a)). The values $w_{k,j}(t) \in \overline{\mathbb{R}}_+$, $j, k \in \mathbb{N}_N$ are the elements of the adjacency matrix $W(t) = [w_{k,j}(t)] \in \mathbb{R}^{N \times N}$ of some communication graph $\mathcal{G}(t) = \{\mathcal{V}, \mathcal{E}(t), W(t)\}$. In view of the results from the previous chapter, we do not only require the outputs $y_k(t)$, $k \in \mathbb{N}_N$ to synchronize but we impose in addition that the synchronous outputs are given by the outputs of some virtual exosystem

$$\dot{\xi}(t) = S\xi(t) \tag{6.3a}$$
$$\eta(t) = R\xi(t) \tag{6.3b}$$

with state vector $\xi(t) \in \mathbb{R}^\nu$ and output $\eta(t) \in \mathbb{R}^q$. That is, we require that for all initial conditions $x_k(0) \in \mathbb{R}^{n_k}$, $z_k(0) \in \mathbb{R}^{m_k}$, $k \in \mathbb{N}_N$, there exist initial conditions $\xi(0) \in \mathbb{R}^\nu$ such that the network consisting of the individual systems (6.1) together with their dynamic coupling controllers (6.2) possesses the property that $(y_k(t) - \eta(t)) \to 0$, $k \in \mathbb{N}_N$ exponentially fast as $t \to \infty$. Without loss of generality, we assume that the pair (R, S) is observable.

The problem we aim at solving is summarized in the following modified version of the Linear Heterogeneous Output Synchronization Problem 5.1:

Problem 6.1 (Linear Heterogeneous Output Synchronization with Virtual Exosystem). *Let the N individual systems (6.1) be modeled by given matrices $A_k \in \mathbb{R}^{n_k \times n_k}$, $B_k \in \mathbb{R}^{n_k \times p_k}$, and $C_k \in \mathbb{R}^{q \times n_k}$ for $k \in \mathbb{N}_N$. Let the communication topology in the couplings (6.2d) be defined by some communication graph $\mathcal{G}(t) = \{\mathcal{V}, \mathcal{E}(t), W(t)\}$ with $|\mathcal{V}| = N$. Let the virtual exosystem (6.3) be given by matrices $S \in \mathbb{R}^{\nu \times \nu}$ and $R \in \mathbb{R}^{q \times \nu}$.*

Find, if possible, matrices $E_k \in \mathbb{R}^{m_k \times m_k}$, $F_k \in \mathbb{R}^{m_k \times r}$, $M_k \in \mathbb{R}^{m_k \times q}$, $G_k \in \mathbb{R}^{p_k \times m_k}$, $H_k \in \mathbb{R}^{p_k \times r}$, $O_k \in \mathbb{R}^{p_k \times q}$, $P_k \in \mathbb{R}^{r \times m_k}$ for $k \in \mathbb{N}_N$, and $Q \in \mathbb{R}^{r \times q}$, such that the pairs

$$\left(\begin{pmatrix} C_k & 0 \end{pmatrix}, \begin{pmatrix} A_k + B_k O_k C_k & B_k G_k \\ M_k C_k & E_k \end{pmatrix} \right), \quad k \in \mathbb{N}_N$$

are detectable and the closed loop of the N systems (6.1) with the dynamic couplings (6.2) is not asymptotically stable and possesses the property that for all initial conditions $x_k(0) \in \mathbb{R}^{n_k}, z_k(0) \in \mathbb{R}^{n_k}$, $k \in \mathbb{N}_N$, there exist initial conditions $\xi(0) \in \mathbb{R}^{\nu}$ such that $(y_k(t) - \eta(t)) \to 0$ exponentially fast as $t \to \infty$ for all $k \in \mathbb{N}_N$ and $(\zeta_k(t) - \zeta_j(t)) \to 0$ exponentially fast as $t \to \infty$ for all $j, k \in \mathbb{N}_N$.

Note that the above problem is stated such that Assumption 5.3 is satisfied for any solution to the problem, i.e., trivial synchronization is excluded in the problem statement.

6.1.2 An Internal Model is Necessary and Sufficient for Linear Output Synchronization

Based on the foregoing discussions, particularly the results from Section 3.3 and 5.1, we give the following result:

Theorem 6.2. *Let $\mathcal{G}(t) = \{\mathcal{V}, \mathcal{E}(t), W(t)\}$ be some communication graph with $|\mathcal{V}| = N > 1$ and Laplacian matrix $L(t)$. Let $A_k \in \mathbb{R}^{n_k \times n_k}$, $B_k \in \mathbb{R}^{n_k \times p_k}$, and $C_k \in \mathbb{R}^{q \times n_k}$ for $k \in \mathbb{N}_N$. Let $S \in \mathbb{R}^{\nu \times \nu}$ and $R \in \mathbb{R}^{q \times \nu}$.*

If the pair (R, S) is observable, $\sigma(S) \subset \mathrm{j}\mathbb{R}$, $\mathcal{G}(t)$ is uniformly connected, there exist matrices $K_k \in \mathbb{R}^{p_k \times n_k}$ and $J_k \in \mathbb{R}^{n_k \times q}$ such that the matrices $A_k + B_k K_k$ and $A_k + J_k C_k$ are Hurwitz for $k \in \mathbb{N}_N$, and there exist matrices $\Pi_k \in \mathbb{R}^{n_k \times \nu}$, $\Lambda_k \in \mathbb{R}^{p_k \times \nu}$, $k \in \mathbb{N}_N$ such that

$$A_k \Pi_k + B_k \Lambda_k = \Pi_k S \tag{6.4a}$$

$$C_k \Pi_k = R \tag{6.4b}$$

for all $k \in \mathbb{N}_N$, then the matrices

$$E_k = \begin{pmatrix} S & 0 \\ \Lambda_k - K_k \Pi_k & A + J_k C_k + B_k K_k \end{pmatrix}, \quad F_k = \begin{pmatrix} I_\nu \\ 0 \end{pmatrix}, \quad M_k = \begin{pmatrix} 0 \\ -J_k \end{pmatrix},$$

$$G_k = \begin{pmatrix} \Lambda_k - K_k \Pi_k & K_k \end{pmatrix}, \qquad P_k = \begin{pmatrix} I_\nu & 0 \end{pmatrix},$$

$H_k = 0$, $O_k = 0$ and $Q = 0$ for all $k \in \mathbb{N}_N$ solve Problem 6.1.

Proof. Define $m_k \triangleq \nu + n_k$ and $z_k(t) \triangleq (\zeta_k(t)^T, \hat{x}_k(t)^T)^T$. Then substituting the matrices defined in the theorem into the couplings (6.2a)–(6.2c) yields

$$\dot{\zeta}_k(t) = S\zeta_k(t) + \delta_k(t) \tag{6.5a}$$
$$\dot{\hat{x}}_k(t) = A_k\hat{x}_k(t) + B_k u_k(t) + J_k(C_k\hat{x}_k(t) - y_k(t)) \tag{6.5b}$$
$$u_k(t) = K_k(\hat{x}_k(t) - \Pi_k\zeta_k) + \Lambda_k\zeta_k \tag{6.5c}$$

for all $k \in \mathbb{N}_N$. Define $e_k(t) \triangleq x_k(t) - \hat{x}_k(t)$ and $\varepsilon_k(t) \triangleq x_k(t) - \Pi_k\zeta_k(t)$. With (6.4a), we then obtain

$$\begin{aligned}\dot{\varepsilon}_k(t) &= (A_k + B_kK_k)x_k(t) - B_kK_ke_k(t) - B_k(K_k\Pi_k - \Lambda_k)\zeta_k(t) - \Pi_kS\zeta_k(t) - \Pi_k\delta_k(t)\\ &= (A_k + B_kK_k)\varepsilon_k(t) - B_kK_ke_k(t) - \Pi_k\delta_k(t),\end{aligned}$$

for $k \in \mathbb{N}_N$. That is, the individual systems with their dynamic coupling controllers in coordinates $\zeta_k(t)$, $e_k(t)$, $\varepsilon_k(t)$ are given as

$$\begin{pmatrix} \dot{\zeta}_k(t) \\ \hline \dot{\varepsilon}_k(t) \\ \dot{e}_k(t) \end{pmatrix} = \left(\begin{array}{c|cc} S & 0 & 0 \\ \hline 0 & A_k + B_kK_k & -B_kK_k \\ 0 & 0 & A_k + J_kC_k \end{array}\right) \begin{pmatrix} \zeta_k(t) \\ \hline \varepsilon_k(t) \\ e_k(t) \end{pmatrix} + \begin{pmatrix} I_\nu \\ \hline -\Pi_k \\ 0 \end{pmatrix} \delta_k(t)$$

$$y_k(t) = \left(\begin{array}{c|cc} R & C_k & 0 \end{array}\right) \begin{pmatrix} \zeta_k(t) \\ \hline \varepsilon_k(t) \\ e_k(t) \end{pmatrix}$$

for all $k \in \mathbb{N}_N$, where we used (6.4b) to obtain the output equation. These systems are detectable by observability of (R, S) and not asymptotically stable since $\sigma(S) \in \mathrm{j}\mathbb{R}$. The dynamics for $e_k(t)$ and $\varepsilon_k(t)$ are governed by an asymptotically stable LTI system driven by the input $\delta_k(t)$. The dynamics for $\zeta_k(t)$, $k \in \mathbb{N}_N$ with static diffusive couplings (6.2d) satisfy the conditions of Theorem 3.22, i.e., $(\zeta_k(t) - \zeta_j(t)) \to 0$ exponentially fast as $t \to \infty$ for all $j, k \in \mathbb{N}_N$ and thus $\delta_k(t) \to 0$ exponentially fast as $t \to \infty$ for all $k \in \mathbb{N}_N$. Thus, by Lemma B.1, $e_k(t) \to 0$ and $\varepsilon_k(t) \to 0$ exponentially fast as $t \to \infty$ for all $k \in \mathbb{N}_N$. The claim then follows from the observation that $\varepsilon_k(t) = 0$ and $\zeta_k(t) = \zeta_j(t)$ implies $y_k(t) = R\zeta_k(t) = R\zeta_j(t) = y_j(t)$. □

It is worth giving some remarks about the dynamic coupling controller (6.5) proposed in the above theorem. In fact, (6.5) is a particular case of the general dynamic coupling controller (6.2) posed in the problem statement. The specific structure of the dynamic coupling controller (6.5) is illustrated in the block diagram depicted in Figure 6.1. The three blocks in the block diagram represent the individual system, a Luenberger observer for the individual system, and a copy, i.e., an internal model, of the virtual exosystem.

We first discuss the local part of the couplings, i.e., the dynamics for $\delta_k(t) \equiv 0$. Note that the input $u_k(t)$ produced by the dynamic coupling controller consists of two parts: $u_k(t) = u_{k,1}(t) + u_{k,2}(t)$ with $u_{k,1}(t) = \Lambda_k\zeta_k(t)$ and $u_{k,2}(t) = K_k(\hat{x}_k(t) - \Pi_k\zeta_k(t))$. The first part $u_{k,1}(t)$ is a feedforward part which guarantees invariance of the local steady state set $\{(x_k, \hat{x}_k, \zeta_k) \in \mathbb{R}^n \times \mathbb{R}^n \times \mathbb{R}^\nu | x_k = \hat{x}_k = \Pi_k\zeta_k\}$. The second part $u_{k,2}(t)$ is a feedback part which renders the local steady state set attractive. In fact, with $\delta_k(t) \equiv 0$, the feedforward control $u_{k,1}(t)$ corresponds exactly to the controller given in the proof of Lemma 5.7 (modulo the observer), where equivalence of solvability of the implicit

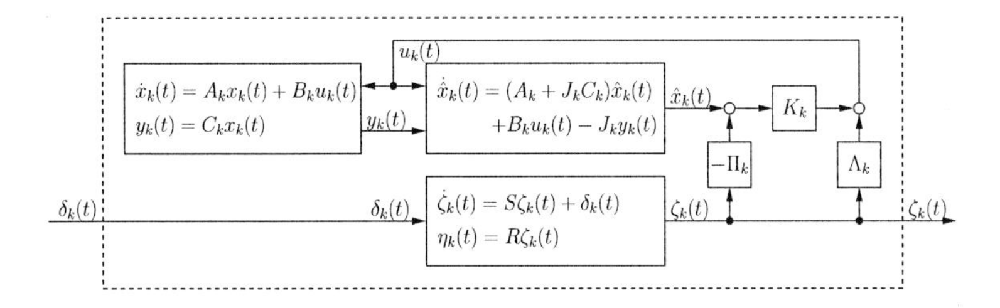

Figure 6.1: Block diagram of an individual system with dynamic coupling controller.

and the explicit internal model conditions has been demonstrated. Existence of such a control is guaranteed by the results from the previous chapter, in particular Corollary 5.8. By Theorem 6.2 above, we know that imposing stabilizability and detectability of the individual systems on top of the conditions of Corollary 5.8 is sufficient to guarantee existence of a control that renders the local state set attractive. In fact, in absence of couplings over the network, i.e., for $\delta_k(t) = 0$, the individual systems with dynamic coupling controllers can be viewed as an exosystem tracked by the individual system with the help of a Luenberger observer. This structure is well-known in linear output regulation (see Francis, 1977, Knobloch et al., 1993).

If $\delta_k(t) \not\equiv 0$, the couplings between the individual systems in the network with their local dynamic coupling controllers yield additional loops on top of the local control loops depicted in Figure 6.1 and discussed above. However, since the dynamics for $\zeta_k(t)$ are not influenced by the remaining local dynamics, from what is 'seen' at the outputs $\zeta_k(t)$, the individual systems look like they were all identical to the virtual exosystem. Thus synchronization of the virtual exosystems is guaranteed by the results presented in Chapter 3.

In summary, output synchronization in the network (6.1) is achieved using a hierarchical strategy. On the local level of the hierarchy, *tracking controllers* ensure that the system output tracks the output of an internal model of the virtual exosystem. On the network level of the hierarchy, a *consensus protocol* based on static diffusive couplings ensures synchronization of the internal models. The combination of the two yields synchronization of the network.

This hierarchical approach comes with the benefit of robustness to individual systems failure. In fact, since the internal models of the virtual systems are synchronized and synchronization of the network is achieved by local tracking controllers, there is no feedback from the individual system to the network. That is, if one of the systems behaves unexpectedly, it may not synchronize with the remaining group, but it will not destroy synchronization in the remaining group neither.

Combining Corollary 5.8 and Theorem 6.2 we obtain the following simple but important corollary:

Corollary 6.3. *Let $\mathcal{G}(t) = \{\mathcal{V}, \mathcal{E}(t), W(t)\}$ be some communication graph with $|\mathcal{V}| = N > 1$ and Laplacian matrix $L(t)$. Let $A_k \in \mathbb{R}^{n_k \times n_k}$, $B_k \in \mathbb{R}^{n_k \times p_k}$, and $C_k \in \mathbb{R}^{q \times n_k}$ for $k \in \mathbb{N}_N$. Let $S \in \mathbb{R}^{\nu \times \nu}$ and $R \in \mathbb{R}^{q \times \nu}$.*

Assume the pairs (C_k, A_k), $k \in \mathbb{N}_N$ are detectable,the pairs (A_k, B_k), $k \in \mathbb{N}_N$ are stabilizable, the pair (R, S) is observable, $\sigma(S) \subset \mathrm{j}\mathbb{R}$, and $\mathcal{G}(t)$ is uniformly connected.

Then Problem 6.1 admits a solution if and only if for all $k \in \mathbb{N}_N$ there exist matrices $\Pi_k \in \mathbb{R}^{n_k \times \nu}$, $\Lambda_k \in \mathbb{R}^{p_k \times \nu}$ that solve (6.4).

Proof. Necessity was shown in Corollary 5.8. Sufficiency is a direct consequence of Theorem 6.2. □

6.1.3 Discussion

To facilitate the understanding of the different assumptions and conditions involved in Theorem 6.2 and Corollary 6.3, we will compare these results to a result proposed in Scardovi and Sepulchre (2008, 2009) for synchronization in networks of identical linear systems stated as follows:

Theorem 6.4. *Let $\mathcal{G}(t) = \{\mathcal{V}, \mathcal{E}(t), W(t)\}$ be some communication graph with $|\mathcal{V}| = N > 1$ and Laplacian matrix $L(t)$. Let $A_k = A \in \mathbb{R}^{n \times n}$, $B_k = B \in \mathbb{R}^{n \times p}$, and $C_k = C \in \mathbb{R}^{q \times n}$ for all $k \in \mathbb{N}_N$.*

Assume the pair (C, A) is detectable and the pair (A, B) is stabilizable, $\sigma(A) \subset \overline{\mathbb{R}}_-$, and $\mathcal{G}(t)$ is uniformly connected.

Then the dynamic couplings

$$\dot{\eta}_k(t) = (A + BK)\eta_k(t) - \sum_{j=1}^{N} w_{k,j}(t)(\eta_k(t) - \eta_j(t) + \hat{x}_j(t) - \hat{x}_k(t)) \tag{6.6a}$$

$$\dot{\hat{x}}_k(t) = A\hat{x}_k(t) + Bu_k(t) + J(C\hat{x}_k(t) - y_k(t)) \tag{6.6b}$$

$$u_k(t) = K\eta_k(t) \tag{6.6c}$$

with $K \in \mathbb{R}^{p \times n}$ such that $(A + BK)$ is Hurwitz and $J \in \mathbb{R}^{n \times q}$ such that $(A + JC)$ is Hurwitz ensures that for any initial conditions $x_k(0) \in \mathbb{R}^n$, $\eta_k(0) \in \mathbb{R}^n$, and $\hat{x}_k(0) \in \mathbb{R}^n$, $k \in \mathbb{N}_N$, there exists some $x_0 \in \mathbb{R}^n$ such that $x_k(t) - e^{At}x_0 \to 0$ exponentially fast as $t \to \infty$ for all $k \in \mathbb{N}_N$.

The proof is similar to the proof of Theorem 6.2. In fact, with the change of coordinates $\zeta_k(t) = \hat{x}_k(t) - \eta_k(t)$, (6.6a) becomes

$$\dot{\zeta}_k(t) = A\zeta_k(t) + J(C\hat{x}_k(t) - y_k(t)) - \sum_{j=1}^{N} w_{k,j}(t)(\zeta_k(t) - \zeta_j(t))$$

and (6.6c) yields $u_k(t) = K(\hat{x}_k(t) - \zeta_k(t))$. Thus, the coupling dynamics (6.6) are almost identical to those proposed in Theorem 6.2 with $\Pi_k = I_n$, $\Lambda_k = 0$ and the internal model of the virtual exosystem replaced by a copy of the individual system itself. The only difference is the observer error appearing in the dynamics for $\zeta_k(t)$, which does not affect convergence properties.

It may appear surprising that the couplings (6.6) proposed by Scardovi and Sepulchre (2009) include an internal model of the individual system itself. Since all individual systems are supposed to admit identical dynamical models, in contrast to the situation considered in Theorem 6.2, no additional internal model is needed to guarantee existence of an appropriate steady state for the global network. There must thus be another motivation for the internal model of the individual system embedded in the couplings (6.6). Indeed such a motivation exists and was already mentioned when discussing consensus with static diffusive couplings over time-varying communication topologies in Section 3.3: the result for consensus with static diffusive couplings over uniformly connected communication graphs given in Theorem 3.22 requires that the input matrix B has full column rank, i.e., there exist as many independent inputs as state components. Although this condition was not shown to be necessary, the discussion in Scardovi and Sepulchre (2009) suggests that stronger assumption on the communication graph are needed to achieve consensus with static diffusive couplings if B does not have full column rank. The dynamic couplings (6.6) are thus used to surmount the necessary tradeoff between system complexity and topological complexity we have been faced with in the case of static diffusive couplings.

Just like observers are used to compensate for the requirement that the output matrix C has full row rank (see also Chapter 4), the internal model in (6.6) compensates for the constraint that the input matrix B has full column rank. However, in the latter case, it turns out that the same type of coupling dynamics allows to compensate for the constraint that all individual systems admit identical dynamical models as shown in Theorem 6.2. While in case of identical systems, it is possible to choose static couplings if one imposes stronger connectedness assumptions on the network, this is not true for the case of non-identical systems, where coupling dynamics are needed to impose a synchronous steady state for the network. That is, dynamic couplings are again necessary to synchronize individual systems with increased complexity while maintaining high topological complexity.

It is worth mentioning that even for networks with identical individual systems, Theorem 6.2 is more general than Theorem 6.4, since the latter requires $\sigma(A) \subset \overline{\mathbb{R}}_-$ while there is no such requirement in Theorem 6.2. In fact, in Theorem 6.2, this requirement is replaced by the constraint $\sigma(S) \in \mathrm{j}\mathbb{R}$ and the tracking controller is used to stabilize unstable parts of the individual systems. Thus, in both cases, the network synchronizes to trajectories that grow at most polynomially fast in time.

A common feature of both results given in Theorems 6.2 and 6.4 is that relative controller states are exchanged over the network, i.e., a communication network with the ability to transmit arbitrary information over the network is needed. Heterogeneous output synchronization with the additional constraint of relative output sensing (similar to the results presented in Chapter 4) is studied in Kim et al. (2010). However, the results in Kim et al. (2010) are restricted to constant communication graphs, i.e., the results presented here are more general in terms of admissible communication topologies while it is more demanding in terms of the type of information being communicated between the individual systems.

It was argued in Section 5.1.2, that the virtual exosystem is part of the design, i.e., one can choose the type of outputs the network shall synchronize to. This fact is stressed by the fact that the matrices $S \in \mathbb{R}^{\nu\times\nu}$ and $R \in \mathbb{R}^{q\times\nu}$ are part of the problem data in Theorem 6.2. While this makes the result general and flexible, it may seem slightly counter-intuitive in the context of consensus and synchronization. When synchronizing a group of dynamical systems, one usually expects that the solutions of the synchronized systems are similar to the solutions of the uncoupled systems. This is however not possible if the individual systems have arbitrary non-identical dynamical models. Yet, usually the individual systems in a network are not completely unrelated. Very often, their dynamical models admit parts that are identical or at least similar for all individual systems and the choice of the virtual exosystem in Theorem 6.2 appears naturally. This is illustrated on a simple example in what follows.

6.1.4 Example

As an example, we consider a network of mobile agents which we assume to be modeled as $\ddot{y}_k(t) = v_k(t)$, $k \in \mathbb{N}_N$ where $y_k(t)$ is the position and $\dot{y}_k(t)$ is the velocity of the kth agent. Heterogeneity comes into play by considering different types of actuators yielding individual systems modeled as

$$\begin{aligned}\dot{x}_k(t) &= \left(\begin{array}{cc|c} 0 & 1 & 0 \\ 0 & 0 & \hat{C}_k \\ \hline 0 & \hat{D}_k & \hat{A}_k \end{array}\right) x_k(t) + \left(\begin{array}{c} 0 \\ 0 \\ \hline \hat{B}_k \end{array}\right) u_k(t) \\ y_k(t) &= \left(\begin{array}{cc|c} 1 & 0 & 0 \end{array}\right)\end{aligned}$$

with state $x_k(t) \in \mathbb{R}^{n_k}$, $n_k \geq 2$, input $u_k(t) \in \mathbb{R}^{p_k}$, and actuator matrices $\hat{A}_k \in \mathbb{R}^{\hat{n}_k \times \hat{n}_k}$, $\hat{B}_k \in \mathbb{R}^{\hat{n}_k \times p_k}$, $\hat{C}_k \in \mathbb{R}^{1 \times \hat{n}_k}$, and $\hat{D}_k \in \mathbb{R}^{\hat{n}_k \times 1}$ with $\hat{n}_k = n_k - 2$ for $k \in \mathbb{N}_N$. Note that the actuators for the individual systems are not required to have the same dynamical order. Furthermore, we allow the actuator dynamics to be influenced by the velocity of the agents through the matrices $\hat{D}_k$.

We want to synchronize the output and the velocity of the individual agents and the synchronous solutions shall correspond to the solutions of an agent with an ideal actuator. This determines the choice

$$S = \begin{pmatrix} 0 & 1 \\ 0 & 0 \end{pmatrix}, \quad R = \begin{pmatrix} 1 & 0 \end{pmatrix}$$

for the matrices modeling the virtual exosystem. By Lemma 5.11, we know that the Francis equations (6.4) are solvable if the individual systems do not have invariant zeros at the origin. In the present case this implies solvability of

$$\begin{pmatrix} \hat{A}_k & \hat{B}_k \\ \hat{C}_k & 0 \end{pmatrix} \begin{pmatrix} \pi_{k,32} \\ \lambda_{k,2} \end{pmatrix} = - \begin{pmatrix} \hat{D}_k \\ 0 \end{pmatrix}$$

for $\pi_{k,32} \in \mathbb{R}^{\hat{n}_k \times 1}$ and $\lambda_{k,2} \in \mathbb{R}^{p_k \times 1}$ for all $k \in \mathbb{N}_N$. With these quantities, the solution to the Francis equations (6.4) read

$$\Pi_k = \begin{pmatrix} 1 & 0 \\ 0 & 1 \\ 0 & \pi_{k,32} \end{pmatrix}, \quad \Lambda_k = \begin{pmatrix} 0 & \lambda_{k,2} \end{pmatrix}.$$

If $\hat{D}_k = 0$, we have $\pi_{k,32} = 0$ and $\lambda_{k,2} = 0$, i.e., $\Lambda_k = 0$. This is explained by the fact that, if $\hat{D}_k = 0$, the individual system with $u_k(t) \equiv 0$ admits an invariant subspace with the property that the dynamics restricted to that subspace corresponds the the dynamics of the virtual exosystem, i.e., no internal model is needed to guarantee existence of the desired steady state of the individual system with dynamic coupling controller.

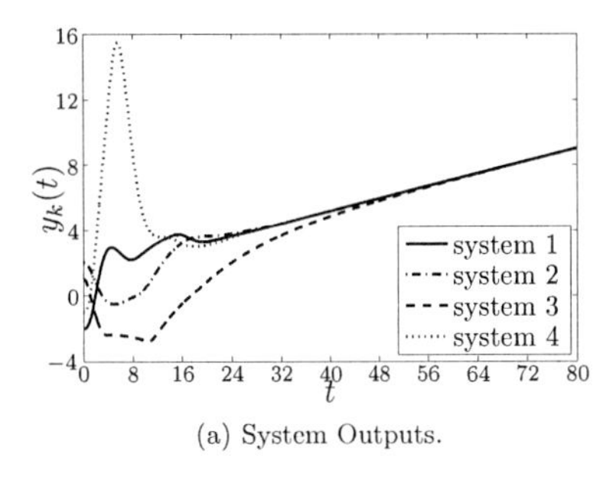

(a) System Outputs.

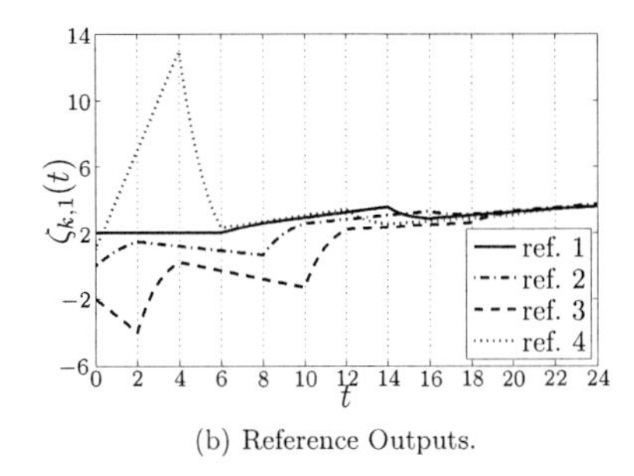

(b) Reference Outputs.

Figure 6.2: Evolution of the system outputs $y_k(t)$, $k \in \mathbb{N}_4$ (a) and the corresponding internal reference signals $\zeta_{k,1}(t)$, $k \in \mathbb{N}_4$ (b) of the closed loop double integrator dynamics with different dynamic actuators. In (b), vertical lines indicate the time instances when changes in the interconnection topology occur.

In a simulation study, we consider a group of four individual systems with $\hat{n}_k = 1$ and $p_k = 1$ for $k \in \mathbb{N}_4$ and parameters $\{\hat{A}_k, \hat{B}_k, \hat{C}_k, \hat{D}_k\}$ chosen as $\{-1, 1, 1, 0\}$, $\{-10, 2, 1, 0\}$, $\{-2, 1, 1, -10\}$, and $\{-2, 1, 1, -1\}$ respectively, i.e., the first system has a slow independent actuator, the second system has a fast independent actuator, the third system's actuator is strongly influenced by the corresponding agent's velocity, and the fourth system's actuator is mildly influenced by the corresponding agent's velocity. The feedback gains K_k and the observer gains J_k, $k \in \mathbb{N}_4$ are chosen using an LQG design for each individual system independently. The time-varying communication topology is assumed periodic with period $T = 8$ with

$$\mathcal{E}(\kappa T + t) = \begin{cases} \{(v_1, v_2)\}, & \text{for } t \in \left[0, \frac{T}{4}\right), \\ \{(v_2, v_3)\}, & \text{for } t \in \left[\frac{T}{4}, \frac{T}{2}\right), \\ \{(v_3, v_4)\}, & \text{for } t \in \left[\frac{T}{2}, \frac{3T}{4}\right), \\ \{(v_4, v_1)\}, & \text{for } t \in \left[\frac{3T}{4}, T\right) \end{cases}$$

for any $\kappa \in \mathbb{N} \bigcup \{0\}$ and $w_{k,j}(t) \in \{0, 1\}$, $i, j \in \mathbb{N}_4$. Initial conditions are chosen randomly.

Figure 6.2(a) shows the system responses $y_k(t)$, $k \in \mathbb{N}_4$. Figure 6.2(b) shows the corresponding controller states $\zeta_{k,1}$, $k \in \mathbb{N}_4$ which are exponentially synchronized with the approach described in Theorem 3.22. It can be seen in Figure 6.2(b) that the controllers synchronize quickly, i.e., the individual systems evolve independently from each other most of the time. The difference between the outputs depicted in Figure 6.2(a) and the references in Figure 6.2(b) are due to observer and tracking errors as well as the different actuator dynamics of the individual systems.

With this example, we conclude the linear case. We showed that solvability of the Francis equations is necessary and sufficient for solvability of the Linear Heterogeneous Output Synchronization Problem 6.1 under mild technical assumptions. We cannot hope for such a general result in case of nonlinear networks. However, some of the ideas used for linear systems will be useful for synchronization of nonlinear oscillators discussed in what follows.

6.2 Synchronization of Nonlinear Oscillators

6.2.1 Problem Setup

The problem considered here is a particular case of the nonlinear heterogeneous output synchronization problem introduced in Section 5.2.1. We consider a network of N exponentially stable oscillators modeled as

$$\dot{x}_k(t) = f_k(x_k(t)) + b_k u_k(t) \tag{6.7}$$

with state $x_k(t) \in \mathbb{R}^{n_k}$, input $u_k(t) \in \mathbb{R}$ and $f_k : \mathbb{R}^{n_k} \to \mathbb{R}^{n_k}$ locally Lipschitz for $k \in \mathbb{N}_N$. The reason for assuming the input to enter the dynamics linearly is mainly for ease of presentation of the results below. It is not completely obvious how to define synchronization in the above network in a meaningful way. Thus, we will define the outputs to be synchronized later on, after some preliminary discussions. We assume that, in absence of exogenous signals, i.e., for $u_k(t) \equiv 0$, all of the above systems admit a particular solution $x_k(t) = \gamma_k(t)$ which is defined for all $t \in \mathbb{R}$, non-constant, and periodic with period T_k, i.e., $\gamma_k(t) = \gamma_k(t+T_k)$ for all $t \in \mathbb{R}$. We denote the corresponding *periodic orbit* as $\Gamma_k \triangleq \gamma_k(\mathbb{R}) \subset \mathbb{R}^{n_k}$, i.e., the periodic orbit Γ_k is the locus of the periodic solution $\gamma_k(\cdot)$. The frequency of the periodic solution $\gamma_k(t)$ is defined in the usual way as $\omega_k \triangleq \frac{2\pi}{T_k}$. We are interested in the particular case when all solutions of (6.7) with $u_k(t) \equiv 0$ starting in a neighborhood of Γ_k converge to a solution contained in Γ_k exponentially fast. In that case, we call Γ_k an exponentially stable limit cycle, formally defined below.

Definition 6.5 (Exponentially Stable Limit Cycle, Asymptotic Phase). *Let $f_k : \mathbb{R}^{n_k} \to \mathbb{R}^{n_k}$ locally Lipschitz and $x_k(t) = \gamma_k(t)$ a periodic solution to (6.7) with $u_k(t) \equiv 0$ of frequency ω_k and periodic orbit $\Gamma_k \triangleq \gamma_k(\mathbb{R})$.*

The periodic orbit $\Gamma_k \subset \mathbb{R}^{n_k}$ is an exponentially stable limit cycle *for (6.7) with $u_k(t) \equiv 0$, if there exists some neighborhood $\mathcal{U}_k \subset \mathbb{R}^{n_k}$ of Γ_k, constants $\mu, M \in \mathbb{R}_+$, and a map $\theta_k : \mathcal{U}_k \to \mathbb{S}^1$ with the property that*

$$\left\| x_k(t) - \gamma_k\left(t + \frac{1}{\omega_k}\theta_k(x_k(0))\right)\right\| \leq Me^{-\mu t}\left\| x_k(0) - \gamma_k\left(\frac{1}{\omega_k}\theta_k(x_k(0))\right)\right\|$$

holds along solutions $x_k(t)$ to (6.7) uniformly in initial conditions $x_k(0) \in \mathcal{U}_k$ and $t \in \overline{\mathbb{R}}_+$.

Given any $x_k \in \mathcal{U}_k$, the value $\varphi_k = \theta_k(x_k)$ is called the asymptotic phase *of x_k with respect to the periodic solution $\gamma_k(\cdot)$.*

Exponential stability of limit cycles is a common assumption in the analysis of oscillatory systems (see Hoppensteadt and Izhikevich, 1997, Izhikevich and Ermentrout, 2008). This assumption guarantees that the asymptotic phase exists and the map $\theta_k(\cdot)$ is well defined in a neighborhood of the periodic orbit and is a key assumption for many more results. Beyond those technical reasons, there is a systems theoretic reason to assume exponential stability of the limit cycles. Namely, without this particular property, also commonly referred to as hyperbolicity (see Fenichel, 1972), the limit cycle is non-robust in the sense that arbitrarily small perturbations may be enough for a bifurcation to occur and the limit cycle to disappear.

The concept of asymptotic phase will be important in what follows and is worth being discussed in some more detail. To this end, we define the phase of some point $x_k \in \Gamma_k$, i.e., a point on the limit cycle, as the unique value $\varphi_k \in \mathbb{S}^1$ satisfying $x_k = \gamma_k\left(\frac{1}{\omega_k}\varphi_k\right)$. Note that with this definition, the phase of a point on the limit cycle is identical to its asymptotic phase. Since all solutions to (6.7) with $u_k(t) \equiv 0$, which start in a neighborhood of the limit cycle, converge to a particular periodic solution, it is natural to define the asymptotic phase of those solutions as the phase of the limiting periodic solution. This is exactly the asymptotic phase defined above. With theses definitions, the phase and asymptotic phase satisfy the differential equation $\dot{\varphi}_k(t) = \omega_k$. It should be noted that the phase and asymptotic phase are defined relative to the solution $\gamma_k(\cdot)$, which is unique modulo time shifts. That is, the choice of a particular solution $\gamma_k(\cdot)$ determines the point on the limit cycle which has zero phase.

Synchronization in networks of oscillators is usually understood as synchronization of the oscillators' phases. That is, in the above network, we would say that the oscillators are synchronized if $\varphi_k(t) = \theta_k(x_k(t)) = \theta_j(x_j(t)) = \varphi_j(t)$ holds identically for all times $t \in \mathbb{R}$ for all $j, k \in \mathbb{N}_N$. We know from the internal model principle discussed in Chapter 5 that dynamic controllers are required for this to be possible, unless $\omega_k = \omega_j$ for all $j, k \in \mathbb{N}_N$.

In fact, for every individual oscillator, a dynamic controller needs to be designed which embeds an internal model of some common virtual exosystem in the local closed loop in such a way that the closed loop possesses a steady state with the property that the steady state dynamics are characterized by the virtual exosystem dynamics. We explained earlier that the virtual exosystem is a degree of freedom in the design procedure, i.e., one can choose the type of synchronous trajectories by appropriately choosing the virtual exosystem. However, since we are interested in synchronization of *oscillators* in the present case, we want to maintain the oscillatory behavior of the uncoupled systems in the coupled network and in particular in steady state, i.e., once the systems are synchronized. Therefore, we are interested in solutions with a virtual exosystem that exhibits oscillatory behavior, i.e., we impose the virtual exosystem as $\dot{\eta}(t) = \hat{\omega}$, $\eta(t) \in \mathbb{S}^1$ for some target frequency $\hat{\omega} \in \mathbb{R}_+$, which is the simplest model for an oscillator. We denote the period of this virtual exosystem as $\hat{T} = \frac{2\pi}{\hat{\omega}}$.

The most proximate solution then seems to be the one which does not change the limit cycles, i.e., the steady state *locus*, of the individual systems, but uniformly speeds up or slows down the oscillators on their limit cycles such that they all oscillate with the common frequency $\hat{\omega}$ of the virtual exosystem. Yet, this cannot be achieved in general without further assumptions on the individual systems, since this would require the possibility to design a control which yields a constant influence tangent to the limit cycle and which has no effect transverse to the limit cycle. Thus we relax the problem in two ways. The first relaxation concerns the steady state locus of the individual oscillators. Instead of requiring the limit cycle of the controlled system to coincide with the limit cycle $\Gamma_k \subset \mathbb{R}^{n_k}$ of the open loop system, we allow the controller to modify the steady state locus yielding a controlled limit cycle $\hat{\Gamma}_k \subset \mathbb{R}^{n_k}$. We only require $\hat{\Gamma}_k$ to be contained in the domain of the phase map $\theta_k(\cdot)$ and to satisfy $\theta_k(\hat{\Gamma}_k) = \mathbb{S}^1$. The synchronized network will then admit periodic solutions $\hat{\gamma}_k : \mathbb{R} \to \hat{\Gamma}_k$, $k \in \mathbb{N}_N$ of period $\hat{T}$ and we can define controlled asymptotic phase maps $\hat{\theta}_k : \hat{\mathcal{U}}_k \to \mathbb{S}^1$ in some neighborhood $\hat{\mathcal{U}}_k$ of $\hat{\Gamma}_k$ using Definition 6.5. The second relaxation concerns the question when we consider a group of oscillators to be

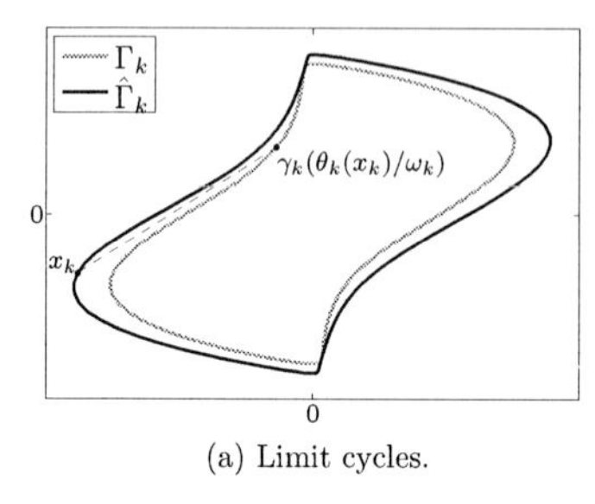

(a) Limit cycles.

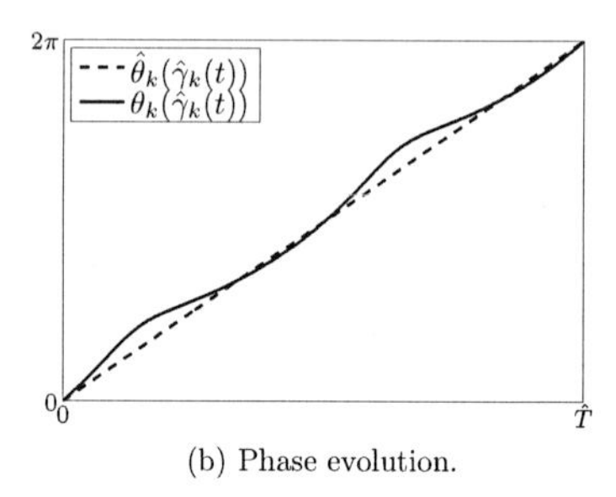

(b) Phase evolution.

Figure 6.3: Unforced and forced limit cycles Γ_k and $\hat{\Gamma}_k$ for a Van der Pol oscillator (a) with evolution of the phases $\hat{\theta}_k(\hat{\gamma}_k(t))$ and $\theta_k(\hat{\gamma}_k(t))$ (b) for the forced periodic solution $\hat{\gamma}_k(t)$ defined relative to the unforced periodic solution $\gamma_k(t)$ and relative to the forced periodic solution itself respectively.

synchronized. Instead of requiring $\varphi_k(t) = \varphi_j(t)$ for all times $t \in \mathbb{R}$ and all $j, k \in \mathbb{N}_N$, we only require this condition to hold true for all $t \in \mathbb{R}$ such that $\varphi_k(t) = 0$ for some $k \in \mathbb{N}_N$. If we choose the controlled periodic solutions $\hat{\gamma}_k(\cdot)$ such that $\theta_k(\hat{\gamma}_k(0)) = 0,\ k \in \mathbb{N}_N$, this corresponds to the condition $\hat{\varphi}_k(t) = \hat{\theta}_k(x_k(t)) = \hat{\theta}_j(x_j(t)) = \hat{\varphi}_j(t)$ identically for all $t \in \mathbb{R}$ and all $j, k \in \mathbb{N}_N$.

To illustrate these relaxations, suppose we are interested in synchronizing an ensemble of analogue metronomes and assume the phase is defined such that the metronomes produce a click whenever the phase is zero. Then the above definition means that the metronomes, if they are synchronized, produce clicks simultaneously, but we do not require the pendulums of the metronomes to move exactly synchronously and we allow the pendulums to move differently for the controlled system than for the open loop system. An example of a limit cycle Γ_k compared to a controlled limit cycle $\hat{\Gamma}_k$ is given in Figure 6.2(a). The asymptotic phase $\hat{\varphi}_k(t) = \theta_k(\hat{\gamma}_k(t))$ of the controlled periodic solution $\hat{\gamma}_k(t)$ defined relative to the unforced periodic solution $\gamma_k(t)$ is exemplarily depicted in Figure 6.2(b).

With the above problem statement and in view of the internal model principle presented in Chapter 5, the first step in designing the dynamic coupling controllers for the network (6.7) is to find dynamic controllers that ensure existence of limit cycles $\hat{\Gamma}_k$ and asymptotic phase maps $\hat{\theta}_k : \hat{\mathcal{U}}_k \to \mathbb{S}^1$ defined in a neighborhood $\hat{\mathcal{U}}_k$ of $\hat{\Gamma}_k$ such that in the local closed loop systems $\frac{\partial \hat{\theta}_k(x_k)}{\partial x_k} f(x_k) + bu_k = \hat{\omega}$ holds for all $k \in \mathbb{N}_N$. That is, the dynamic coupling controllers need to be such that the dynamics of the controlled individual systems on $\hat{\Gamma}_k$ correspond to the virtual exosystem dynamics. In the linear case, the corresponding steady state dynamics was imposed using a standard tracking controller. For the case of nonlinear oscillators, a tracking controller is hard to design. In particular, it would require a priori knowledge of the controlled limit cycles $\hat{\Gamma}_k$, $k \in \mathbb{N}_N$. For this reason, we use a different approach here. Namely, we use entrainment of oscillators to guarantee the desired steady state behavior of the individual oscillators.

6.2.2 Entrainment of Oscillators by Small Periodic Forcing

Exponentially stable limit cycles possess some useful properties that will be partially exploited in what follows. In particular, exponentially stable limit cycles are structurally robust. In fact any exponentially stable limit cycle Γ_k of some oscillator (6.7) is an attractive normally hyperbolic invariant manifold, i.e., convergence to the manifold is exponentially fast and contraction on the manifold is slower than convergence to the manifold (see Fenichel, 1972 or Hoppensteadt and Izhikevich, 1997, Section 4.3). This implies that Γ_k persists under small perturbations. We will be interested in the case when small periodic forcing yields a perturbed limit cycle $\hat{\Gamma}_k$ for the oscillator (6.7) with the period of the forcing.

A key tool in this analysis is the reduction of the oscillator dynamics (6.7) to a simple one-dimensional phase model. By definition of the maps $\theta_k : \mathcal{U}_k \to \mathbb{S}^1$ (see Definition 6.5), we know that for $u_k(t) \equiv 0$ the asymptotic phase $\varphi_k(t) = \theta_k(x_k(t))$ satisfies

$$\dot{\varphi}_k(t) = \omega_k$$

for all $k \in \mathbb{N}_N$ (see Hoppensteadt and Izhikevich, 1997, Izhikevich, 2007, Izhikevich and Ermentrout, 2008, Pikovsky et al., 2001). If we let $u_k(t) = \varepsilon_k \cos(\hat{\omega}t)$ for some small parameter $\varepsilon_k \in \mathbb{R}_+$ and some frequency $\hat{\omega} \in \mathbb{R}_+$, we obtain a perturbed version of the phase model as

$$\dot{\varphi}_k(t) = \omega_k + \varepsilon_k \langle q_k(\varphi_k(t)), b_k \rangle \cos(\hat{\omega}t) + \mathcal{O}(\varepsilon_k^2) \tag{6.8}$$

where $\langle \cdot, \cdot \rangle$ denotes the standard inner product in $\mathbb{R}^n$. The map $q_k : \mathbb{S}^1 \to \mathbb{S}^1$ is the *infinitesimal phase response curve* (iPRC) for the kth oscillator. It accounts for effects of the input tangent to the limit cycle while the higher order terms in ε_k account for perturbations of the locus of the limit cycle. For any $k \in \mathbb{N}_N$, the iPRC $q_k : \mathbb{S}^1 \to \mathbb{S}^1$ can be obtained as the unique periodic solution to the adjoint problem

$$\frac{\mathrm{d}q_k(\varphi_k)}{\mathrm{d}\varphi_k} = -\frac{1}{\omega_k} \{Df_k(\gamma_k(\varphi_k/\omega_k))\}^T q_k(\varphi_k)$$

satisfying the normalization condition $\langle q_k(0), f(\gamma_k(0)) \rangle = 1$. (Malkin's Theorem, see Izhikevich, 2007). Defining the phase difference $\chi_k(t) \triangleq \varphi_k(t) - \hat{\omega}t$ and the normalized frequency mismatch $\nu_k \triangleq (\omega_k - \hat{\omega})/\varepsilon_k$ yields

$$\dot{\chi}_k(t) = \varepsilon_k g_k(\chi_k(t), t, \varepsilon_k)$$

with

$$g_k(\chi_k, t, \varepsilon_k) \triangleq \nu + \langle q_k(\chi_k + \hat{\omega}t), b_k \rangle \cos(\hat{\omega}t) + \mathcal{O}(\varepsilon_k)$$

for $k \in \mathbb{N}_N$. Thus, if ν_k and ε_k are small enough, the phase difference $\chi_k(t)$ varies slowly compared to the forcing phase $\hat{\omega}t$ and averaging can be applied to obtain approximations $\chi_k(t) \approx \vartheta_k(t)$ with

$$\dot{\vartheta}_k(t) = \varepsilon_k \overline{g}_k(\vartheta_k(t)) + \mathcal{O}(\varepsilon_k^2), \tag{6.9}$$

where $\overline{g}_k$ is defined as

$$\overline{g}_k(\vartheta_k) \triangleq \frac{1}{\hat{T}} \int_0^{\hat{T}} g(\vartheta_k, t, 0) \mathrm{d}t$$

for all $k \in \mathbb{N}_N$. If (6.9) has a hyperbolic equilibrium, then $\chi_k(t) = \vartheta_k(t) + \mathcal{O}(\varepsilon_k)$ uniformly in $t \in \overline{\mathbb{R}}_+$, i.e., the approximation is valid for all times $t \in \overline{\mathbb{R}}_+$. Details can be found in Hoppensteadt and Izhikevich (1997, Chapter 9).

The maps $\overline{g}_k$ are 2π-periodic in ϑ_k. Its precise form depends on the vector fields $f_k(x_k)$ and on the forcing signal. Using the fact that $u_k(t) = \varepsilon_k \cos(\hat{\omega} t)$ if harmonic forcing is applied, we can however predict the general form of $\overline{g}_k$ with the help of the following lemma:

Lemma 6.6. *Let $Q : \mathbb{R} \to \mathbb{R}$ be a 2π-periodic map. Then there exist $a \in \mathbb{R}$ and $\hat{\vartheta} \in \mathbb{S}^1$ such that*

$$\frac{1}{\hat{T}} \int_0^{\hat{T}} Q(\vartheta + \hat{\omega} t) \cos(\hat{\omega} t) \mathrm{d}t = a \cos(\vartheta + \hat{\vartheta}).$$

Proof. Consider the change of variables $\varphi = \vartheta - \hat{\omega} t$ to obtain

$$\begin{aligned}
\frac{1}{\hat{T}} \int_0^{\hat{T}} Q(\vartheta + \hat{\omega} t) \cos(\hat{\omega} t) \mathrm{d}t &= \frac{1}{2\pi} \int_{\vartheta}^{\vartheta + 2\pi} Q(\varphi) \cos(\varphi - \vartheta) \mathrm{d}\varphi \\
&= \frac{1}{2\pi} \int_0^{2\pi} Q(\varphi) \cos(\varphi - \vartheta) \mathrm{d}\varphi.
\end{aligned}$$

Using the identity $\cos(\varphi - \vartheta) = \cos(\varphi)\cos(\vartheta) + \sin(\varphi)\sin(\vartheta)$, terms in ϑ can be pulled out of the integral and the lemma follows. □

Defining $Q_k(\varphi_k) \triangleq \langle q_k(\varphi_k), b_k \rangle$, $k \in \mathbb{N}_N$, we know by the above lemma that there exist gains $a_k \in \mathbb{R}$ and phase shifts $\hat{\vartheta}_k \in \mathbb{S}^1$, $k \in \mathbb{N}_N$ such that

$$\overline{g}_k(\vartheta_k) = \nu_k + a_k \cos(\vartheta_k + \hat{\vartheta}_k)$$

for all $k \in \mathbb{N}_N$. As a consequence, for ε_k and ν_k small enough, the averaged phase difference equation (6.9) possesses a unique exponentially stable equilibrium, i.e., the phase difference $\chi_k(t) = \varphi_k(t) - \hat{\omega} t$ will become constant on average and the asymptotic behavior of the oscillators (6.7) is entirely determined by the forcing signal.

Thus, in particular, the forced oscillators (6.7) admit periodic solutions $\hat{\gamma}_k(\cdot)$ and limit cycles $\hat{\Gamma}_k = \hat{\gamma}_k(\mathbb{R})$ and it is possible to define an asymptotic phase map $\hat{\theta}_k : \hat{\mathcal{U}}_k \to \mathbb{S}^1$ in some neighborhood $\hat{\mathcal{U}}_k$ of $\hat{\Gamma}_k$ relative to the solution $\hat{\gamma}_k(\cdot)$, such that $\hat{\varphi}_k(t) = \hat{\theta}_k(\hat{\gamma}_k(t))$ satisfies $\dot{\hat{\varphi}}_k(t) = \hat{\omega}$ for all $t \in \mathbb{R}$. With a constant phase shift, i.e., $u_k = \varepsilon_k \cos(\hat{w} t + \hat{\xi}_k)$ for some constant $\hat{\xi}_k \in \mathbb{S}^1$, we can moreover achieve that $\hat{\theta}_k(\hat{\gamma}_k(t)) = 0$ if and only if $\theta_k(\hat{\gamma}_k(t)) = 0$. The example depicted in Figure 6.3 shows the limit cycles and phase evolutions of a forced Van der Pol oscillator $\dot{x}_{k,1}(t) = -3(x_{k,2}^2(t) - 1)x_{k,1}(t) - x_{k,2}(t) + u_k(t)$, $\dot{x}_{k,2}(t) = x_{k,1}(t)$ with $u(t) = \cos(\omega_k t + \hat{\xi}_k)$ where $\omega_k \approx 0.709$ is the natural frequency of the unforced oscillator and $\hat{\xi}_k \approx \pi/3$.

The above result can be interpreted in terms of the internal model principle from Chapter 5 as follows: The virtual exosystem is defined as $\dot{\xi}(t) = \hat{\omega}$, $\eta(t) = \xi(t)$ with $\xi(t) \in \mathbb{S}^1$. Define the maps $\pi_k : \mathbb{S}^1 \to \hat{\Gamma}_k$ and $\lambda_k : \mathbb{S}_1 \to \mathbb{R}$ as $\pi_k(\xi) \triangleq \hat{\gamma}_k(\xi/\hat{\omega})$ and $\lambda_k(\xi) = \varepsilon_k \cos(\xi + \hat{\xi}_k)$. If we consider $\hat{\theta}_k(x_k)$ as the oscillators output, the FBI equations read

$$\begin{aligned}
\frac{\partial \hat{\gamma}_k(\xi/\hat{\omega})}{\partial \xi} \hat{\omega} &= f_k(\hat{\gamma}_k(\xi/\hat{\omega})) + b\varepsilon_k \cos(\xi + \hat{\xi}_k) \\
\xi &= \hat{\theta}_k(\hat{\gamma}_k(\xi/\hat{\omega}))
\end{aligned}$$

for all $k \in \mathbb{N}_N$. The equations are trivially satisfied. Substituting $\xi/\hat{\omega} = t$ in the first equation yields exactly the defining equation (6.7) for the oscillator with input $u_k(t) = \varepsilon \cos(\hat{\omega}t + \hat{\xi}_k)$ evaluated along the forced solution $\hat{\gamma}(t)$ which was obtained using exactly that input. The second equation reduces to the definition of the phase map $\hat{\theta}_k(x_k)$ for $x_k \in \hat{\Gamma}_k$. It turns out in this particular case, that the dynamic controller

$$\dot{z}_k(t) = \hat{\omega} \tag{6.10a}$$

$$u_k(t) = \varepsilon_k \cos(z_k(t) + \hat{\xi}_k) \tag{6.10b}$$

with state $z_k \in \mathbb{S}^1$, which is obtained as a copy of the virtual exosystem with $u_k(t) = \lambda_k(z_k(t))$, not only renders the set $\{(x_k, z_k) \in \mathbb{R}^{n_k} \times \mathbb{S}^1 | x_k = \pi_k(z_k), z_k \in \mathbb{S}^1\}$ invariant but it also renders this set locally exponentially attractive for all $k \in \mathbb{N}_N$. That is, using the stability and robustness properties of exponentially stable limit cycles mentioned at the outset of this section, a feedforward control is enough to achieve invariance *and* attractivity of the controlled steady state of the oscillators. Comparing the controller (6.10) to the linear tracking controllers (6.5), the control $u_k(t)$ defined in (6.10b) corresponds to the feedforward part $\Lambda_k \zeta_k(t)$ in the control (6.5c). The feedback part $K_k(\hat{x}_k(t) - \Pi_k \zeta_k(t))$ from (6.5c) has no counterpart in (6.10b) since attractivity of the steady state set defined above comes for free in the case of exponentially stable limit cycles.

It must be noted that the above convergence results are local results due to different reasons. If the unforced systems (6.7) with $u_k(t) \equiv 0$ possess non-trivial invariant subsets of their state spaces other than the limit cycles $\Gamma_k \subset \mathbb{R}^n$, $k \in \mathbb{N}_N$, it is likely that the forced systems possess invariant sets other than the forced limit cycle $\hat{\Gamma}_k \subset \mathbb{R}^n$, $k \in \mathbb{N}_N$. Furthermore, the averaged phase difference equation (6.9) always possesses an unstable equilibrium, i.e., there exist sets $\tilde{\Gamma}_k$ (possibly identical to $\hat{\Gamma}_k$) and maps $\tilde{\pi}_k : \mathbb{S}^1 \to \tilde{\Gamma}_k$ for all $k \in \mathbb{N}_N$ such that the set $\tilde{\Gamma}_k^* \triangleq \{(x_k, z_k) \in \tilde{\Gamma}_k \times \mathbb{S}^1 | x_k = \tilde{\pi}_k(z_k), z_k \in \mathbb{S}^1\}$ is an unstable limit cycle for the closed loop system. If the unforced oscillator possesses a unique limit cycle and all other invariant sets are unstable, almost all solutions of the closed loop system will converge the forced limit cycle $\hat{\Gamma}_k$.

6.2.3 Phase Synchronization through Entrainment by a Consensus Input

Based on the discussion above and guided by the result for linear systems from Section 6.1, we can state the following result:

Theorem 6.7. *Let $\mathcal{G}(t) = \{\mathcal{V}, \mathcal{E}(t), W(t)\}$ be some communication graph with $|\mathcal{V}| = N > 1$ and Laplacian matrix $L(t)$. Let $f_k : \mathbb{R}^{n_k} \to \mathbb{R}^{n_k}$ be such that the oscillators (6.7) with $u_k \equiv 0$ possess unique, exponentially stable limit cycles $\Gamma_k \subset \mathbb{R}^{n_k}$ of period $T_k \in \mathbb{R}_+$ and frequency $\omega_k \in \mathbb{R}_+$ and no other stable invariant sets for all $k \in \mathbb{N}_N$ and let $b_k \in \mathbb{R}^n$ such that $\langle q_k(\varphi_k), b_k \rangle \not\equiv 0$ where $q_k : \mathbb{S}^1 \to \mathbb{S}^1$ is the iPRC of the kth system.*

If $\max_{k \in \mathbb{N}_N} \omega_k - \min_{k \in \mathbb{N}_N} \omega_k$ is sufficiently small, then there exist $\hat{\omega} \in \mathbb{R}_+$, $\varepsilon_k \in \mathbb{R}_+$,

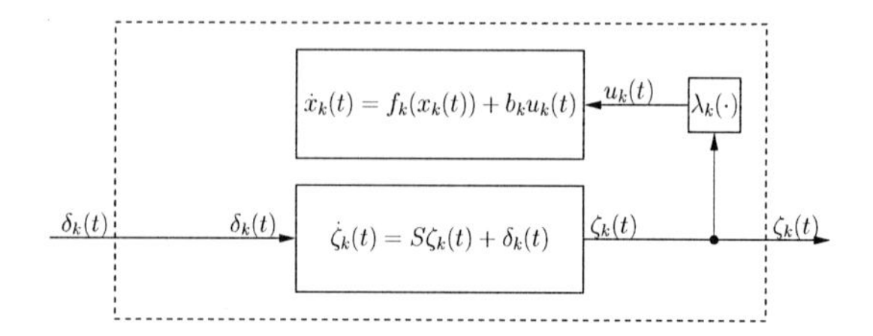

Figure 6.4: Block diagram of individual oscillator with dynamic coupling controller.

and $\hat{\xi}_k \in \mathbb{S}^1$*, for all* $k \in \mathbb{N}_N$ *such that the dynamic couplings*

$$\dot{\zeta}_k(t) = \begin{pmatrix} 0 & -\hat{\omega} \\ \hat{\omega} & 0 \end{pmatrix} \zeta_k(t) + \delta_k(t) \tag{6.11a}$$

$$u_k(t) = \begin{cases} \dfrac{\varepsilon_k}{\|\zeta_k(t)\|} \left(\cos\left(\hat{\xi}_k\right), \sin\left(\hat{\xi}_k\right)\right) \zeta_k(t), & \|\zeta_k(t)\| \neq 0 \\ 0, & \|\zeta_k(t)\| = 0 \end{cases} \tag{6.11b}$$

$$\delta_k(t) = -\sum_{j=1}^{N} w_{k,j}(t)(\zeta_k(t) - \zeta_j(t)) \tag{6.11c}$$

ensure synchronization of the network (6.7) *for almost all initial conditions* $x_k(0) \in \mathbb{R}^{n_k}$*,* $\zeta_k(0) \in \mathbb{R}^2$*,* $k \in \mathbb{N}_N$*.*

Proof. By the discussion of Section 6.2.2, $u_k(t) = \varepsilon_k \cos(\hat{\omega}t + \hat{\xi}_k)$ yields entrainment with exponential convergence rate for almost all initial conditions.

Furthermore, by Theorem 3.22, $u_k(t) - \varepsilon_k \cos(\hat{\omega}t + \hat{\xi}_k + t_0) \to 0$ exponentially fast as $t \to \infty$ for almost all initial conditions $\zeta_k(0) \in \mathbb{R}^2$ (in fact all initial conditions that do not yield consensus of the systems (6.11a) at the origin). We can assume without loss of generality that $t_0 = 0$.

Since the forced limit cycles $\hat{\Gamma}_k$, $k \in \mathbb{N}_N$ are normally hyperbolic attractive manifolds (see Fenichel, 1972), it follows that solutions converge to arbitrary small neighborhoods of $\hat{\Gamma}_k$, $k \in \mathbb{N}_N$ in finite time. Thus, in particular, they converge in finite time to the domain of validity of the linearization about the periodic solution. □

The structure of the dynamic coupling controller (6.11) is depicted in Figure 6.4. It is very similar to the structure of the dynamic coupling controller (6.5) used in the linear case in Section 6.1.2 and depicted in Figure 6.1. The only structural difference is that the feedback term $\hat{x}_k(t) - \Pi_k\zeta_k(t)$ is missing in the couplings (6.11). The reason for this is that attractivity of the steady state set is guaranteed by feedforward control only, as explained in Section 6.2.2. Thus, once the internal models are synchronized, synchronization is achieved by a stimulus or forcing signal which is common to all individual systems. Similar mechanisms are popular, e.g., in synchronization problems for populations of neurons (see Bulsara et al., 1996).

Compared to the controller (6.10) proposed in Section 6.2.2 to achieve entrainment, the dynamic coupling controllers (6.11) proposed in the above theorem looks a bit more

Table 6.1: Comparison of Theorem 6.7 to existing results for synchronization of nonlinear oscillators.

	Slotine and Wang (2004)	**Stan and Sepulchre (2007)**	**Theorem 6.7**
Idea	Contraction Analysis	Dissipativity	Consensus and Entrainment
Couplings	static, diffusive		dynamic
Graph	balanced, connected		directed, uniformly connected
Systems	identical, partial contraction property	identical, incrementally passive	non-identical, exponentially stable limit cycles

complicated at first glance. However, the two are closely related and (6.11) with $\delta_k(t) \equiv 0$ is obtained from (6.10) by adding a second state $r_k(t) \in \mathbb{R}$ with dynamics $\dot{r}_k(t) = 0$ and using the transformation $\zeta_k(t) = (r_k(t)\cos(z_k(t)), r_k(t)\sin(z_k(t)))^T$ for $r_k(t) = \|\zeta_k(t)\| \neq 0$. We choose the slightly more complicated representation (6.11) for the dynamic coupling controllers in order to be able to apply Theorem 3.22. With this choice, the steady state locus of the local closed loops of the oscillators (6.7) with the dynamic coupling controllers (6.11) can be defined as $\{(x_k, \zeta_k) \in \mathbb{R}^{n_k} \times \mathbb{R}^2 | x_k = \hat{\theta}_k(\arg(\zeta_{k,1} + \mathrm{j}\zeta_{k,2}) + \hat{\xi}_k), \|\zeta_k\| > 0\}$. The steady state dynamics are thus defined by an internal model of a linear harmonic oscillator and synchronization is achieved by synchronizing these internal models, exactly as in the linear case considered in Section 6.1.

In order to apply the control (6.11), it is necessary to know the precise phase shifts $\hat{\xi}_k$, $k \in \mathbb{N}_N$ for all oscillators in the group. This generally requires either precise knowledge of the oscillator models or the possibility to determine the phase shifts experimentally. In many cases, neither of the two is possible. Furthermore, if the phase shifts $\hat{\xi}_k$ are determined incorrectly, this will result in synchronization errors that do not decay to zero. To overcome this problem, it is a straightforward extension of the dynamic coupling control (6.11) to determine the phase shifts $\hat{\xi}_k$ adaptively by a slow feedback loop added on top of the controller (6.11). Using again the example of the metronomes introduced in the problem setup, the simple idea is to adjust the phase shift $\hat{\xi}_k$ by a small amount every time the metronome produces a click, such that $\arg\big((\cos(\hat{\xi}_k), \mathrm{j}\sin(\hat{\xi}_k))\zeta_k(t)\big)$ becomes closer to zero. We will not further pursue this approach here.

6.2.4 Discussion

The result presented in Theorem 6.7 is, to the author's best knowledge, the first result for synchronization of nonlinear oscillators over uniformly connected communication graphs. Existing results like Slotine and Wang (2004), Stan and Sepulchre (2007), Wang and Slotine (2005) usually allow for synchronization over fixed and balanced communication topologies using static diffusive couplings. In addition the only requirement imposed by Theorem 6.7 on the individual systems is existence of an exponentially stable limit cycle. It has been argued in the problem statement already, that this assumption is often reasonable and does not impose an important restriction. Since limit cycles that are not exponentially stable are not structurally robust, they are rarely observed in physical systems. As a matter of fact, the conditions imposed on the individual systems by Slotine and Wang (2004), Stan and Sepulchre (2007), Wang and Slotine (2005) imply exponential stability of the limit cycles under consideration. The results mentioned above are compared to the result proposed in Theorem 6.7 in Table 6.1.

Similarly to the linear case, where we showed that the dynamic couplings (6.6) proposed by Scardovi and Sepulchre (2009) can be used with very small changes to allow for synchronization among *non-identical* systems, the coupling dynamics (6.11) proposed in Theorem 6.7 allow for synchronization among *non-identical* oscillators. The permissible heterogeneity is limited in the approach presented here since the difference in the unforced frequencies of the oscillators needs to be small. In addition, the precise meaning of small depends on the specific oscillators. In fact, it boils down to the question to what range of frequencies a given oscillator can be entrained by harmonic forcing. That question has no generic answer. Despite this restriction, the proposed dynamic couplings are much more flexible than static diffusive couplings. In fact, it is a consequence of the internal model principle discussed in Chapter 5, that it is impossible to synchronize oscillators with different frequencies by means of static diffusive couplings. Thus, the coupling *dynamics* clearly allow for solutions with increased system and topological complexity.

In contrast to most of the existing approaches, that are typically based on static diffusive couplings, we use diffusive couplings only to synchronize the linear internal models of some virtual exosystem embedded in the local closed loop systems by means of dynamic coupling controllers. This allows to apply Theorem 3.22 and thereby, we can guarantee exponential synchronization of the internal models over uniformly connected communication graphs. Synchronization among the oscillators (in the sense defined in the problem statement) is achieved on a local level of the hierarchical approach using entrainment by small periodic forcing with the forcing signal produced by the synchronized internal models.

On the one hand, it has been argued already that the entrainment taking place on the local level of hierarchy is purely a feedforward effect. Static diffusive couplings on the other hand are purely a feedback mechanism. We know from Chapter 5 that feedforward control is generally necessary for synchronization among non-identical systems and illustrated in this chapter why it is also sufficient in case of exponentially stable oscillators. But besides the implication for heterogeneous networks, the fact that only feedforward control is used has an important influence on the sensing and communication capabilities that are needed for synchronization.

While synchronization by means of static diffusive couplings requires only fairly simple sensing capabilities, namely only relative sensors are needed, synchronization by means of entrainment does not require any sensors at all, simply because it is a feedforward mechanism. This might be an important advantage especially in the case of nonlinear oscillators, where it is often difficult or even impossible to sense relative or absolute states. Simple improvements, like an adaptive control to determine the phase shifts $\hat{\xi}_k$ can be implemented based on sensors that merely detect zero phase times. While it might be convenient to do without or with very simple sensors, there is of course some downside: The approach of Theorem 6.7 requires some communication network to exchange relative controller states. However, the amount of information, that needs to be communicated over the network is small, since we only require uniform connectedness of the communica-

Table 6.2: Parameter values for Van der Pol oscillators considered in simulations.

k	1	2	3	4	5
μ_k	2	3	4	5	6
ω_k	0.8235	0.7092	0.6158	0.5411	0.4810
T_k	7.630	8.859	10.20	11.61	13.06
ε_k	1	1	1	1	1
$\hat{\xi}_k$	4.303	4.827	5.176	5.480	5.814

tion graph. Thus, our approach requires no sensing and simple communication networks while approaches based on static diffusive couplings require fairly simple sensing capabilities and no high-level communication network. Both setups are of potential use in different applications.

Finally, we comment on the rate of convergence. Theorem 6.7 requires the gains ε_k, $k \in \mathbb{N}_N$ to be sufficiently small. However, the values ε_k directly influence the convergence speed as seen from the averaged phase difference equation (6.9). Thus, it is not possible to give any convergence bounds. The admissible values for ε_k, $k \in \mathbb{N}_N$ crucially depend on the specific oscillators. As a rule of thumb, one can say that the faster solutions converge to the limit cycle of the unforced oscillator, the larger the gains ε_k can be tuned.

6.2.5 Example

As an example, we consider a network of Van der Pol oscillators modeled as

$$\dot{x}_k(t) = \begin{pmatrix} x_{k,2}(t) + \mu_k \left(x_{k,1}(t) - \frac{1}{3}x_{k,1}^3(t)\right) \\ -x_{k,1}(t) \end{pmatrix} + \begin{pmatrix} 0 \\ 1 \end{pmatrix} u_k(t)$$

with state vector $x_k(t) \in \mathbb{R}^2$, input $u_k(t) \in \mathbb{R}$, and parameter $\mu_k \in \mathbb{R}_+$ for $k \in \mathbb{N}_N$. The Van der Pol oscillator is a popular member of the class of Liénard systems (see Liénard, 1928, Slight et al., 2008) for which existence and uniqueness of the limit cycle can be shown. Furthermore it is well-known that the unique limit cycle of the Van der Pol oscillator is exponentially stable (see Grasman et al., 2005). The parameter μ_k has a strong influence on the steady state behavior of the Van der Pol oscillator. For small values of μ, the Van der Pol oscillator is very similar to a linear harmonic oscillator while for large values of μ, relaxation oscillations are observed. Moreover, for increasing values of μ, the frequency of oscillation of the unforced Van der Pol oscillator decreases.

To pose the synchronization problem among a group of Van der Pol oscillators with possibly non-identical parameter values μ_k, $k \in \mathbb{N}_N$, we need to define the zero phase points, i.e., the points on the respective limit cycles that should be visited by the synchronized oscillators simultaneously. We define the phase to be zero if $x_{k,2}(t) = 0$ with positive derivative for all $k \in \mathbb{N}_N$. Of course, any other choice would be possible. This choice merely affects the phase shifts $\hat{\xi}_k \in \mathbb{S}^1$, $k \in \mathbb{N}_N$.

Below, we will give simulation results for a group of 5 Van der Pol oscillators with parameter values given in Table 6.2. We choose the frequency $\hat{\omega}$ of the virtual exosystem, i.e., the target frequency for the synchronized network to be identical to the unforced frequency of the third oscillator, i.e., $\hat{\omega} = \omega_3$. Note that the frequency mismatch between

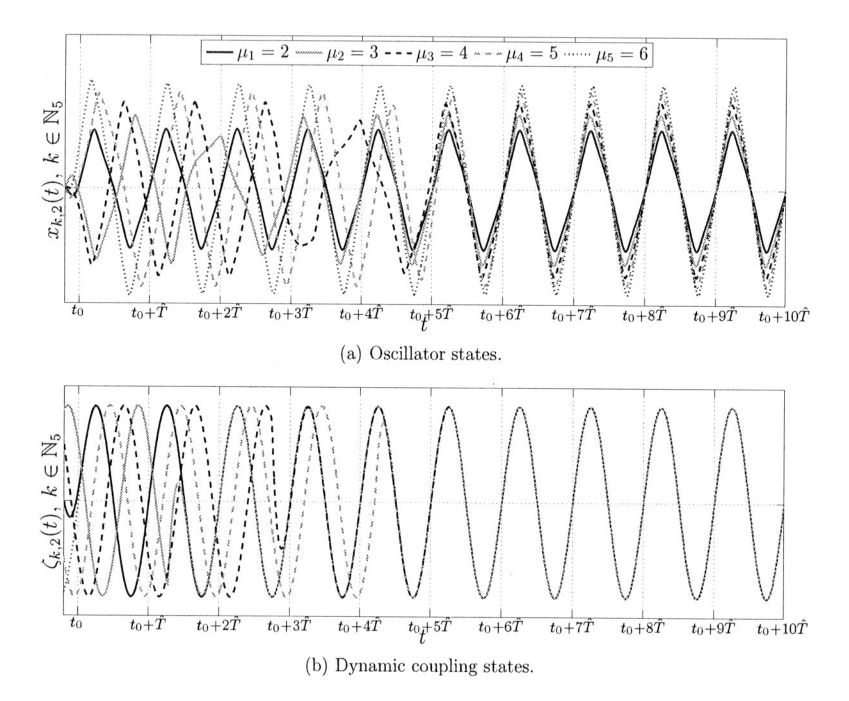

(a) Oscillator states.

(b) Dynamic coupling states.

Figure 6.5: Synchronization of non-identical Van der Pol oscillators (a) by entrainment with a synchronized harmonic forcing signal (b).

the individual oscillators is important. The frequency of the fastest oscillator is more than 70% bigger compared to the frequency of the slowest oscillator. The parameters ε_k have been chosen as $\varepsilon_k = 1$, $k \in \mathbb{N}_N$ and the parameters $\hat{\xi}_k$, $k \in \mathbb{N}_N$ have been determined in simulations as given in Table 6.2.

The communication graph used to synchronize the internal models is assumed, similarly to the graph considered in Section 6.1.4, periodic with period $T = 75$ with

$$\mathcal{E}(\kappa T + t) = \begin{cases} \{(v_5, v_1)\}, & \text{for } t \in \left[0, \frac{T}{5}\right), \\ \{(v_1, v_2)\}, & \text{for } t \in \left[\frac{T}{5}, \frac{2T}{5}\right), \\ \{(v_2, v_3)\}, & \text{for } t \in \left[\frac{2T}{5}, \frac{3T}{5}\right), \\ \{(v_3, v_4)\}, & \text{for } t \in \left[\frac{3T}{5}, \frac{4T}{5}\right), \\ \{(v_4, v_5)\}, & \text{for } t \in \left[\frac{4T}{5}, T\right) \end{cases}$$

for any $\kappa \in \mathbb{N} \bigcup \{0\}$ and $w_{k,j}(t) \in \{0, 1\}$, $i, j \in \mathbb{N}_4$. That is, the graph contains exactly one edge at any fixed time t and the union graph over one period T is a directed cycle.

Simulation results are depicted in Figure 6.5. Figure 6.5(b) shows the state components $\zeta_{k,2}(t)$, $k \in \mathbb{N}_5$ of the dynamic coupling controller. After only a few periods of oscillation, they converge to a common harmonic forcing signal. Figure 6.5(a) shows the oscillator state components $x_{k,2}(t)$, $k \in \mathbb{N}_5$. The synchronization objective, namely $x_{k,2}(t) = 0$ to occur simultaneously for all $k \in \mathbb{N}_5$, is obviously achieved after some periods of oscillation. However, due to heterogeneity of the network, the oscillator state trajectories are clearly not synchronized identically for all times, as expected.

6.3 Summary

In this chapter, we derived sufficient conditions for synchronization among non-identical systems based on the ideas and the understanding gained from the discussion of the necessary conditions given in Chapter 5.

The first part of this chapter was devoted to synchronization of non-identical linear systems. We were able to give a very general result in that case (see Wieland and Allgöwer, 2009a). Namely, if the individual systems are stabilizable and detectable, the internal model principle is necessary and sufficient for synchronization of heterogeneous linear networks. In addition, the internal model requirement given in Corollary 5.8 can always be satisfied for some linear virtual exosystem. The admissible virtual exosystems are only constrained to not contain poles at the zeros of the individual systems. Furthermore, the virtual exosystems must not contain poles in the open right-half complex plane for consensus over uniformly connected communication networks to be applicable, i.e., we need to exclude exponentially unstable synchronous trajectories. The solution proposed for linear systems consists of two levels of hierarchy: on the network level of the hierarchy, consensus results from Chapter 3 are used to synchronize internal models while on the local level of the hierarchy, dynamic tracking controllers are implemented to make the individual systems track the synchronized internal models. It was argued that coupling dynamics are needed to cope with both uniformly connected communication graphs and non-identical system dynamics.

In the second part of this chapter, a similar hierarchical approach was used for synchronization of non-identical nonlinear oscillators. The network level of the hierarchy was exactly the same as in the linear case, namely we used results from Chapter 3 to synchronize internal models of some linear virtual exosystem. For the case of synchronization of oscillators, the virtual exosystem was modeled as a linear harmonic oscillator. On the local level of the hierarchy, the tracking controllers used in the linear case have been replaced by entrainment of oscillators by small periodic forcing. We showed that, using structural robustness properties of exponentially stable limit cycles, it is possible to generate an attractive invariant steady state set for the individual oscillators by simply acting upon the oscillators with a small periodic forcing signal, i.e., a purely feedforward mechanism. With this result, we were able to propose a synchronization scheme for nonlinear oscillators representing several improvements over existing results: the only assumption on the communication graph is uniform connectedness, the only assumption on the individual oscillators is exponential stability of the limit cycles, and the individual oscillators may be non-identical, including non-identical unforced frequencies.

Chapter 7

Conclusions

7.1 Summary and Discussion

We started this thesis with the observation that consensus and synchronization are everywhere. Numerous phenomena from various scientific and engineering disciplines fundamentally rely on consensus and synchronization. In view of this, the objective of the present thesis was to contribute to an improved systems theoretic understanding of some mechanisms underlying consensus and synchronization. Diffusive couplings are among the most important of these mechanisms. They are commonly considered in both consensus and synchronization problems. For this reason, they constituted the main theme in this thesis. In Chapters 2 and 3, we focused on classical *static* diffusive couplings (with minor modifications in case of consensus with a leader considered in Section 3.4). In Chapters 4 and 6, we presented diffusive couplings extended by *dynamic* compensators in order to overcome various limitations inherent to static diffusive couplings and thereby allowing for increased system and topological complexity of the network.

Together with link complexity, system and topological complexity form the three main dimensions of complexity for interconnected systems, yielding the complexity cube introduced in Section 1.1. In fact, consensus and synchronization problems for networks of individual systems pursue very similar objectives and, in both cases, these objectives are typically achieved by means of diffusive couplings, as e.g. in the Kuramoto model for synchronization and the Vicsek model for consensus introduced in Section 1.2. Yet, the two problems are usually classified very differently in the complexity cube. Consensus is often focused on topological complexity, in particular communication constraints, while synchronization problems are usually more concerned with system complexity. To understand and possible surmount the tradeoff between topological complexity on the one hand and system complexity on the other hand, we started by specifying appropriate models for the communication topology and the individual system dynamics. Specifically, in this thesis, we considered the case where the communication topology is modeled by communication graphs – described in detail in Chapter 2 – and the individual systems are modeled as ODEs.

The topological complexity was expressed in terms of assumptions and preconditions imposed on the communication graph. Weaker assumptions on the communication graph lead to increased topological complexity of the network. The best we were able to deal with were time-varying, directed, and uniformly connected communication graphs (Section 3.3 and Chapters 5, 6). The remaining results required constant, directed, and connected graphs (Chapters 3, 4). In some instances, stronger results could be obtained by enforcing balanced or undirected graphs instead of directed graphs. In order to achieve consensus in networks of arbitrary, identical LTI systems over constant communication graphs, graph connectedness is usually not even enough. This is reflected by the conditions provided in Sections 3.2 and 3.4, which generally depend on the spectrum of the Laplacian matrix of the communication graph. As a general insight from the analysis of consensus with static diffusive couplings performed in Chapter 3, we found that weak assumptions on the communication graph yield constraints on the individual system dynamics and vice versa. Specifically, consensus over uniformly connected communication graphs requires full control of the individual system's states, i.e., the input to be of the same dimension as the state, and constrains the individual systems to have no exponentially unstable modes. If the communication graph is assumed constant and connected, with known bounds on the spectrum of the Laplacian matrix, the requirement of having full control of the individual system's states may be dropped and consensus can be achieved for individual systems modeled by arbitrary, identical LTI systems, independent of their stability properties.

All results from Chapter 3 provide statements about consensus or synchronization by means of relative *state feedback*. This can be interpreted as the constraint of having full information of the individual system's states, i.e., the output to be of the same dimension as the system's state. In classical feedback stabilization problems, similar constraints are easily removed using observers, yielding a *dynamic*, stabilizing controller. We showed in Chapter 4 that the problem of designing observers to be used in conjunction with diffusive couplings is non-trivial. Time-varying communication graphs and the constraint of relative output sensing that is often relevant in consensus problems yield important additional challenges. We proposed solutions based on full order and reduced order unknown input observers for the case of constant communication graphs. The resulting couplings were interpreted as *dynamic* generalization of the static diffusive couplings considered in Chapter 3.

The symmetry inherent to static diffusive couplings, resulting from the fact that only relative information is used, yields an additional important constraint on the dynamical models of the individual systems in the network. Namely, the models of the individual systems need to be similar in a specific sense. This constraint was shown for the case of a network of identical linear systems with a leader of potentially different dynamics in Section 3.4. In Chapter 5, the problem of consensus and synchronization in heterogeneous networks was addressed for linear and nonlinear networks with arbitrary communication topologies. We obtained an internal model principle for synchronization, saying that a network of non-identical systems can synchronize only if it is possible to design local dynamic controllers such that the individual systems together with their local dynamic controllers embed an *internal model* of a common *virtual exosystem*. This shows that coupling dynamics are needed to synchronize heterogeneous networks. We derived explicit consentability and synchronizability conditions imposed on the individual system dynamics. These conditions take the form of linear matrix equations in case of linear networks and nonlinear partial differential equations in case of nonlinear networks. They relate the individual system dynamics to the virtual exosystem dynamics. More precisely, these conditions express the fact that the individual systems possess controlled invariant subsets of their state spaces with the property that the closed loop dynamics restricted to these subsets can be controlled to be a copy, i.e., an internal model, of the virtual exosystem dynamics. The key observation was that any couplings, static or dynamic, that yield synchronization in a network of non-identical systems needs to be designed in such a way that the coupled network possesses a synchronous steady state set, i.e., an invariant attractive subset of the state space of the complete network with the property that solutions contained in this set yield identically synchronous outputs. The internal model conditions enforce existence of an invariant set possessing this property. Since attractivity of this set is not included in the internal model conditions, they are necessary but in general not sufficient for synchronizability of the network.

We established close links between the internal model principle for synchronization and the classical internal model principle of control theory of Francis and Wonham (1976) as well as the theory of linear and nonlinear output regulation. In fact, the internal model conditions for synchronization correspond to those known from output regulation with the exosystem replaced by the virtual exosystem. Thus, this thesis connects two important areas of research in systems and control theory, namely consensus and synchronization on the one hand and output regulation on the other hand. In contrast to the classical output regulation problem, there is no autonomous exosystem in synchronization problems. The virtual exosystem merely exists as part of the dynamical models of the individual systems with their local controllers. Since all parts in the network potentially influence any other part in the network, there are multiple feedback loops in the network and, in particular, there is no exosystem with autonomous dynamics. In addition, since we considered the case of time-varying communication topologies, those feedback loops may change over time and the resulting system dynamics are time-varying, i.e., non-autonomous. We showed that the internal model principle remains valid in this configuration, i.e., despite the aforementioned differences, synchronization problems and output regulation problems admit similar necessary solvability conditions.

In Chapter 6, we derived sufficient conditions for the solvability of consensus and synchronization problems for the case of arbitrary linear heterogeneous networks and the case of networks of non-identical exponentially stable oscillators. In both cases, we used static diffusive couplings to synchronize internal models. In the linear network, we used tracking controllers to synchronize the individual systems while in the network of oscillator entrainment by small periodic forcing was used. Both solutions require a high-level communication network to exchange controller states between the individual systems. It was shown that this type of dynamic couplings is also beneficial for networks of identical systems as it allows to relax assumptions and preconditions imposed on the individual system dynamics and the communication graph.

We thus showed how static diffusive couplings impose various constraints and limitations on systems and topological complexity in consensus and synchronization problems. We explained how dynamic couplings are suitable to surmount these limitations and to handle increased system and topological complexity at the same time. In particular, we demonstrated that dynamic couplings allow for synchronization of heterogeneous networks with high system complexity over uniformly connected communication graphs.

7.2 Outlook

Throughout this thesis, we considered a very specific definition of consensus and synchronization. Namely, we required outputs or states of a group of individual systems to be asymptotically the same. This definition may however be too restrictive in some cases. The discussion of an appropriate problem setup for synchronization of nonlinear oscillators in Section 6.2 supports this argument. A broad definition of synchronization is, e.g., given in Blekhman et al. (1997). Since the internal model principle from Chapter 5 merely relies on the property that the individual systems are uncoupled once the systems are synchronized, it is expected that the results remain valid whenever this specific property holds true in a network. If this property is not satisfied, the general need for an appropriate steady state set in the closed loop system will not change; however, this will not necessarily result in conditions that need to be independently satisfied by the individual systems. A detailed analysis of the internal model principle for broader definitions of synchronization may carry over some understanding of the properties and features of specific notions of synchronization.

Besides restricting our focus to a specific definition of synchronization, we only considered exact solutions to consensus and synchronization problems in the sense that the outputs or states were required to be asymptotically exactly identical. Yet, very often it is enough to guarantee that synchronization errors become asymptotically small. This property is sometimes referred to as practical or approximate synchronization (see Blekhman et al., 1997). Such a type of synchronization implies in particular that couplings do not vanish asymptotically, i.e., the internal model principle does not hold in the form stated in this thesis. In Andrews et al. (2006, 2008), an approximate internal model principle has been proposed in the context of biochemical systems, relating an adaptation error to an internal model mismatch. It may be interesting to carry this idea over to synchronization problems and relate the synchronization error to an internal model mismatch.

Another extension of the results presented in this thesis may concern the coupling mechanism used in the network. It might be interesting to ask what are the distinctive features of different coupling mechanisms and how do these features affect the results presented in this thesis. Besides the mechanism of diffusive couplings considered in this thesis, impulsive couplings and synchronization by a common stimulus are frequently considered.

Impulsive couplings are, e.g., presumed by Peskin (1975) in a model for synchronization of pacemaker cells of the heart and they are standard in many other biological models (see Canavier and Achuthan, 2007, Mauroy and Sepulchre, 2008, Memmesheimer and Timme, 2006, Mirollo and Strogatz, 1990). They have been interpreted in a hybrid systems framework in Goebel et al. (2009). The adaptive determination of the phase shifts discussed in Section 6.2.3 can be considered as an example for an impulsive mechanism. In contrast to diffusive couplings, which require continuous sensing of relative states or outputs, impulsive couplings require sensors that detect specific events. Similar to diffusive couplings, impulsive couplings usually have no effect once the network is synchronized.

Synchronization by a common stimulus is popular, e.g., in populations of neurons (Bulsara et al., 1996, Izhikevich, 2007). Since synchronization by a common stimulus is a feedforward mechanism, no sensors are needed in this mechanism. The relevance of feedforward control for synchronization among non-identical systems has been highlighted in Chapters 5 and 6. In fact, the solution presented in Section 6.2 can be interpreted as synchronization by a common stimulus generated by means of consensus among harmonic oscillators.

Some distinctive features of the different coupling mechanisms are compared in Ta-

Table 7.1: Comparison of different mechanisms leading to consensus and synchronization.

	Diffusive Couplings	**Impulsive Couplings**	**Common Stimulus**
General Mechanism	continuous feedback	impulsive feed-back	feedforward
Sensing Requirements	relative sensing	detection of events	no sensing
Behavior once Synchronized	uncoupled when synchronized	uncoupled when synchronized	persistently forced even when synchronized

ble 7.1. It is expected that these different features play an important role in achievable topological and system complexity of the network. Furthermore, it is interesting to interpret the different coupling mechanisms with respect to the internal model principle derived in this thesis.

The results presented in this thesis may thus serve as a starting point for a sound theoretical understanding of consensus and synchronization in a broader sense. In particular, a profound understanding of the implications and the relevance of different notions of synchronization and different coupling mechanisms may help to establish further links to classical control theory and thereby open up new analysis and design possibilities in consensus and synchronization problems.

Appendix A

Intermediate Results and Proofs for Chapter 2

A.1 Perron-Frobenius Theorem

A famous result crucial in the proofs of the results of Chapter 2 is the Perron-Frobenius Theorem. In the interest of making this thesis self-contained, the Perron-Frobenius Theorem is repeated here together with some definitions.

Definition A.1 (Spectral Radius, cf. Horn and Johnson, 1985, Def. 1.1.4). *For a matrix $W \in \mathbb{R}^{n \times n}$, the spectral radius $\rho(W)$ is the maximum modulus of the eigenvalues of W, i.e., $\rho(W) \triangleq \max_{\lambda \in \sigma(W)} |\lambda|$.*

Definition A.2 (Reducible Matrix, cf. Horn and Johnson, 1985, Def. 6.2.21). *A matrix $W \in \mathbb{R}^{n \times n}$ is said to be* reducible *if either $n = 1$ and $W = 0$ or $n \geq 2$ and there is a permutation matrix $P \in \mathbb{R}^{n \times n}$ and some integer r with $1 \leq r \leq n-1$ such that*

$$P^T W P = \begin{pmatrix} B & C \\ 0 & D \end{pmatrix}$$

where $B \in \mathbb{R}^{r \times r}$, $D \in \mathbb{R}^{(n-r) \times (n-r)}$, $C \in \mathbb{R}^{r \times (n-r)}$, and $0 \in \mathbb{R}^{(n-r) \times r}$ is a zero matrix.

Definition A.3 (Irreducible Matrix, cf. Horn and Johnson, 1985, Def. 6.2.22). *A matrix $W \in \mathbb{R}^{n \times n}$ is said to be* irreducible *if it is not reducible.*

With these definitions, we are ready to state the Perron-Frobenius Theorem (cf. Horn and Johnson, 1985, Theorem 8.4.4).

Theorem A.4 (Perron-Frobenius). *Let $W \in \mathbb{R}^{n \times n}$ and suppose that W is irreducible and element-wise non-negative, then*

(a) the spectral radius $\rho(W)$ of W satisfies $\rho(W) > 0$;

(b) the spectral radius $\rho(W)$ of W is an algebraically (and hence geometrically) simple eigenvalue of W;

(c) There is an element-wise positive vector p such that $Wp = \rho(W)p$.

A.2 Intermediate Results

For the readers convenience, we repeat Assumption 2.2 and Definition 2.10 of independent strongly connected components (iSCC) of a graph, which will be used below.

Assumption 2.2. *The graph $\mathcal{G}(t) = \{\mathcal{V}, \mathcal{E}(t), W(t)\}$ is such that the elements of the adjacency matrix $W(t)$ satisfy $w_{k,j}(t) \in \{0\} \bigcup (\alpha, \infty) \subset \overline{\mathbb{R}}_+$ for some fixed threshold $\alpha \in \mathbb{R}_+$, i.e., $(v_j, v_k) \in \mathcal{E}(t)$ if and only if $w_{k,j}(t) > 0$ for all $k, j \in \mathbb{N}_{|\mathcal{V}|}$.*

Definition 2.10. *At fixed time t, an* independent strongly connected component *(iSCC) of a digraph $\mathcal{G}(t) = \{\mathcal{V}, \mathcal{E}(t), W(t)\}$ is an induced subgraph $\tilde{\mathcal{G}} = \{\tilde{\mathcal{V}}, \tilde{\mathcal{E}}\}$ which is maximal, subject to being strongly connected, and satisfies $(v, \tilde{v}) \notin \mathcal{E}(t)$ for any $v \in \mathcal{V} \setminus \tilde{\mathcal{V}}$ and $\tilde{v} \in \tilde{\mathcal{V}}$. That is $\tilde{G}$ is strongly connected and the unweighted digraph induced by any set $\hat{\mathcal{V}}$ with $\tilde{\mathcal{V}} \subseteq \hat{\mathcal{V}} \subseteq \mathcal{V}$ is strongly connected if and only if $\hat{\mathcal{V}} = \tilde{\mathcal{V}}$. Furthermore, there is no edge in $\mathcal{E}(t)$ with tail outside $\tilde{\mathcal{V}}$ and head in $\tilde{\mathcal{V}}$.*

The following result relates irreducibility to graph connectedness and is obtained from Horn and Johnson (1985, Def. 6.2.11 and Theorems 6.2.14 and 6.2.24):

Lemma A.5. *Let t be fixed time and $\mathcal{G}(t) = \{\mathcal{V}, \mathcal{E}(t), W(t)\}$ a digraph satisfying Assumption 2.2. Let $L(t)$ be the Laplacian matrix of $\mathcal{G}(t)$.*

Then, the adjacency matrix $W(t)$ is irreducible if and only if the Laplacian matrix $L(t)$ is irreducible if and only if $\mathcal{G}(t)$ is strongly connected.

With this fact, the following corollary is readily shown:

Lemma A.6. *Let t be fixed time and $\mathcal{G}(t) = \{\mathcal{V}, \mathcal{E}(t), W(t)\}$ a strongly connected digraph satisfying Assumption 2.2. Let $L(t)$ be the Laplacian matrix of $\mathcal{G}(t)$.*

Then the Laplacian matrix $L(t)$ has a simple eigenvalue 0 and there exists an element-wise positive vector $p \in \mathbb{R}^{|\mathcal{V}|}$ such that $L^T(t)p = 0$.

Proof. By Lemma A.5, $W(t)$ is irreducible. Since $W(t)$ is element-wise non-negative, this implies that the diagonal degree matrix $D(t) = \operatorname{diag}(W(t)1_{|\mathcal{V}|})$ is positive definite. We define the normalized adjacency matrix $\tilde{W} = D^{-1}(t)W(t)$ and the corresponding normalized Laplacian $\tilde{L} = D^{-1}(t)L(t) = I - \tilde{W}$.

All rows of the normalized adjacency matrix $\tilde{W}$ sum up to one, i.e., $\tilde{W}1_{|\mathcal{V}|} = 1_{|\mathcal{V}|}$. As a consequence of the Geršgorin Disk Theorem (Horn and Johnson, 1985, Theorem 6.1.1), the spectral radius of $\tilde{W}$ satisfies $\rho(\tilde{W}) = 1$. By the Perron-Frobenius Theorem A.4, 1 is a simple eigenvalue of $\tilde{W}$ and there exists an element-wise positive vector $\tilde{p} \in \mathbb{R}^{|\mathcal{V}|}$ such that $\tilde{W}^T\tilde{p} = \tilde{p}$. Therefore 0 is a simple eigenvalue of $\tilde{L}^T$ and $\tilde{L}^T\tilde{p} = 0$. Consequently, 0 is a simple eigenvalue of $L(t)$ and $p = D\tilde{p} \in \mathbb{R}^{|\mathcal{V}|}$ is an element-wise positive vector such that $L^T(t)p = 0$. □

Finally, we need a result for irreducibly diagonally dominant matrices due to Taussky.

Definition A.7 (Irreducibly Diagonally Dominant). *Let $L = [l_{i,j}] \in \mathbb{R}^{N\times N}$. We say that L is* irreducibly diagonally dominant *if*

(a) L is irreducible;

(b) L *is diagonally dominant, i.e.,* $|l_{k,k}| \geq \sum_{\substack{j \in \mathbb{N}_N \\ j \neq k}} |l_{k,j}|$ *for all* $k \in \mathbb{N}_N$*;*

(c) *For at least one* $k \in \mathbb{N}_N$*, we have* $|l_{k,k}| > \sum_{\substack{j \in \mathbb{N}_N \\ j \neq k}} |l_{k,j}|$.

Lemma A.8 (Taussky, see Horn and Johnson, 1985, Corollary 6.2.27). *Let* $L \in \mathbb{R}^{N \times N}$ *be irreducibly diagonally dominant. Then* L *is invertible.*

A.3 Proofs

Below we will repeat and prove Theorems 2.12 and 2.13.

A.3.1 Proof of Theorem 2.12

Theorem 2.12. *Let* t *be fixed time and* $\mathcal{G}(t) = \{\mathcal{V}, \mathcal{E}(t), W(t)\}$ *a digraph. Let* $v \in \mathcal{V}$.

Then there exists at least one iSCC $\tilde{\mathcal{G}} = \{\tilde{\mathcal{V}}, \tilde{\mathcal{E}}\}$ *such that either* $v \in \tilde{\mathcal{V}}$ *or* $v \in \mathcal{D}(w, \mathcal{G}(t))$ *for any* $w \in \tilde{\mathcal{V}}$*, denoted as* $v \in \mathcal{D}(\tilde{\mathcal{V}}, \mathcal{G}(t))$.

Proof. We shall prove that any vertex $v \in \mathcal{V}$, either belongs to an iSCC or is descendant of all vertices of at least one iSCC.

Let $\tilde{\mathcal{G}} = \{\tilde{\mathcal{V}}, \tilde{\mathcal{E}}\}$ be an iSCC of $\mathcal{G}(t)$ at time t and assume $w \in \tilde{\mathcal{V}}$ is such that $v \in \mathcal{D}(w)$. Since $w \in \mathcal{D}(u)$ for all $u \in \tilde{\mathcal{V}} \setminus \{w\}$, by transitivity, this implies that $v \in \mathcal{D}(u)$ for all $u \in \tilde{\mathcal{V}}$. Therefore, it suffices to show that v either belongs to an iSCC or is descendant of one vertex contained in some iSCC.

Given a vertex $v \in \mathcal{V}$, we claim that the following algorithm stops after a finite number of steps at v itself, if it belongs to an iSCC, or at a vertex $w \in \mathcal{V}$ belonging to an iSCC with $v \in \mathcal{D}(w)$, if v does not belong to an iSCC. To initialize the algorithm, set $\mathcal{W}_0 = \emptyset$, $v_1 = v$, and $k = 1$.

(S1) If v_k belongs to some iSCC, the algorithm terminates.

(S2) Set $\mathcal{W}_k = \{v_k\} \bigcup \left\{ w \in \bigcap_{l=0}^{k-1} \mathcal{W}_l \,|\, v_k \in \mathcal{D}(w) \wedge w \in \mathcal{D}(v_k) \right\}$.

(S3) Choose $v_{k+1} \in \mathcal{T}_k$ with $\mathcal{T}_k \triangleq \{w \in \mathcal{V} \setminus \mathcal{W}_k | \exists u \in \mathcal{W}_k : (w, u) \in \mathcal{E}(t)\}$.

(S4) Increase k by 1 and continue at Step (S1).

As long as the algorithm does not terminate in Step (S1), we have that $|\mathcal{W}_k| \geq 1$ since $v_k \in \mathcal{W}_k$ by definition. Furthermore, the graph induced by $\mathcal{W}_k$ is strongly connected by definition. Since v_k does not belong to an iSCC, the graph induced by $\mathcal{W}_k$ cannot be an iSCC, i.e., it cannot be independent, which implies that $\mathcal{T}_k \neq \emptyset$. That is, if the algorithm terminates during step k^*, v_{k^*} belongs to an iSCC and either $v_{k^*} = v$ or $v \in \mathcal{D}(v_{k^*})$.

To show that the algorithm terminates in a finite number of steps, we show that $v_{k_1} = v_{k_2}$ if and only if $k_1 = k_2$ by contradiction. We assume without loss of generality that $k_2 > k_1$. First note that $k_2 \neq k_1 + 1$, because $\mathcal{T}_{k_1} \bigcap \mathcal{W}_{k_1} = \emptyset$ by definition. Assume $k_2 > k_1 + 1$ and $v_{k_1} = v_{k_2}$. Then on the one hand $v_{k_2} \in \mathcal{D}(v_{k_2-1})$ and on the other hand $v_{k_2} \in \mathcal{T}_{k_2-1}$ implies that $v_{k_2-1} \in \mathcal{D}(v_{k_2})$. However, this implies $v_{k_2} \in \mathcal{W}_{k_2-1}$, a contradiction to $v_{k_2} \in \mathcal{T}_{k_2-1}$. This shows that the algorithm terminates after at most $|\mathcal{V}|$ steps. ∎

A.3.2 Proof of Theorem 2.13

Theorem 2.13. *Let t be fixed time and $\mathcal{G} = \{\mathcal{V}, \mathcal{E}(t), W(t)\}$ a digraph with $|\mathcal{V}| = N$ vertices, satisfying Assumption 2.2. Assume $\mathcal{G}(t)$ has $c \geq 1$ distinct iSCCs $\tilde{\mathcal{G}}_i = \{\tilde{\mathcal{V}}_i, \tilde{\mathcal{E}}_i\}$, $i \in \mathbb{N}_c$.*

Then, $\dim(\ker(L^T(t))) = c$ and there exist unique (modulo vertex permutations) vectors $p^i \in \mathbb{R}^N, i \in \mathbb{N}_c$ satisfying

$$\begin{aligned} p^i_j &> 0 && \text{if } v_j \in \tilde{\mathcal{V}}_i, \\ p^i_j &= 0 && \text{if } v_j \notin \tilde{\mathcal{V}}_i, \end{aligned} \qquad i \in \mathbb{N}_N$$

element-wise and $(p^i)^T 1_N = 1$, $i \in \mathbb{N}_c$, such that $\ker(L^T(t)) = \operatorname{span}(p^1, \ldots, p^c)$, i.e., the c vectors p^i, $i \in \mathbb{N}_c$ span a non-negative, orthogonal basis of the kernel of the transposed Laplacian matrix $L^T(t)$.

Proof. Let $\tilde{\mathcal{G}}_i = \{\tilde{\mathcal{V}}_i, \tilde{\mathcal{E}}_i\}$, $i \in \mathbb{N}_c$ denote the c iSCCs of $\mathcal{G}(t)$. Let $\tilde{N}_i = |\tilde{\mathcal{V}}_i|$ for $i \in \mathbb{N}_c$, $N_0 = N - \sum_{i=1}^{c} \tilde{N}_i$, and $\mathcal{V}_0 = \mathcal{V} \setminus \bigcup_{i \in \mathbb{N}_c} \tilde{\mathcal{V}}_i$. Without loss of generality, we may assume that

$$L(t) = \left(\begin{array}{ccc|c} \tilde{L}_1 & & 0 & \\ & \ddots & & 0 \\ 0 & & \tilde{L}_c & \\ \hline & -M_0 & & \operatorname{diag}(M_0 \cdot 1_{N_0}) + L_0 \end{array}\right),$$

where $\tilde{L}_i \in \mathbb{R}^{\tilde{N}_i \times \tilde{N}_i}, i \in \mathbb{N}_c$ are the Laplacian matrices corresponding to the iSCCs $\tilde{\mathcal{G}}_i$, $M_0 \in \mathbb{R}^{(N-N_0) \times N_0}$ is an element-wise non-negative matrix, and $L_0 \in \mathbb{R}^{N_0 \times N_0}$ is the Laplacian matrix corresponding to the graph induced by $\mathcal{V}_0$.

By Lemma A.6, the matrices $\tilde{L}_i(t)$ have simple eigenvalues 0 and there exists unique element-wise positive vectors $\tilde{p}^i \in \mathbb{R}^{\tilde{N}_i}$ such that $(\tilde{p}^i)^T 1_{\tilde{N}_i} = 1$ and $\tilde{L}_i^T \tilde{p}^i = 0$. Consequently, the vectors $p^i \in \mathbb{R}^N$, $i \in \mathbb{N}_c$, defined by

$$(p^1, \ldots, p^c) = \left(\begin{array}{ccc} \tilde{p}^1 & & 0 \\ & \ddots & \\ 0 & & \tilde{p}^c \\ \hline 0 & \cdots & 0 \end{array}\right),$$

are pairwise orthogonal and satisfy $(p^i)^T 1_N = 1$ and $L^T(t) p^i = 0$, $i \in \mathbb{N}_c$. To prove Theorem 2.13, it remains to show that the matrix $\operatorname{diag}(M_0 \cdot 1_{N_0}) + L_0 \in \mathbb{R}^{N_0 \times N_0}$ is invertible.

To this end, define $D_0 = \operatorname{diag}(M_0 \cdot 1_{N_0}) \in \mathbb{R}^{N_0 \times N_0}$ and let $\mathcal{G}_0 = \{\mathcal{V}_0, \mathcal{E}_0, W_0\}$ be the graph induced by $\mathcal{V}_0$ with Laplacian L_0. Define a sequence of graphs $\{\mathcal{G}_k = \{\mathcal{V}_k, \mathcal{E}_k, \mathcal{W}_k\}\}$, $k = 0, \ldots, k^*$ as follows: Denote by c_k the number of iSCCs of the graph $\mathcal{G}_k$ and by $\tilde{\mathcal{V}}_{k,i}$, $i \in \mathbb{N}_{c_k}$ the vertex sets of those iSCCs. Define $\mathcal{V}_{k+1} = \mathcal{V}_k \setminus \bigcup_{i \in \mathbb{N}_{c_k}} \tilde{\mathcal{V}}_{k,i}$. If $|\mathcal{V}_{k+1}| > 0$, define $\mathcal{G}_{k+1} = \{\mathcal{V}_{k+1}, \mathcal{E}_{k+1}, \mathcal{W}_{k+1}\}$ to be the graph induced by $\mathcal{V}_{k+1}$. If $|\mathcal{V}_{k+1}| = 0$, the sequence stops and $k^* = k$. Since any graph has at least one iSCC and any iSCC contains at least one vertex, the sequence contains at most $|\mathcal{V}_0|$ elements, i.e., $k^* \leq |\mathcal{V}_0| - 1$.

Without loss of generality, we may assume that, for $k = 0, \dots, k^* - 1$,

$$D_k + L_k = \left(\begin{array}{ccc|c} D_{k,1} + \tilde{L}_{k,1} & & 0 & \\ & \ddots & & 0 \\ 0 & & D_{k,c_k} + \tilde{L}_{k,c_k} & \\ \hline & -M_k & & D_k^* + \operatorname{diag}(M_k \cdot 1_{|\mathcal{V}_{k+1}|}) + L_{k+1} \end{array}\right),$$

where $\tilde{L}_{k,i} \in \mathbb{R}^{|\tilde{\mathcal{V}}_{k,i}| \times |\tilde{\mathcal{V}}_{k,i}|}$, $i \in \mathbb{N}_{c_k}$ are the Laplacian matrices corresponding to the iSCCs $\tilde{\mathcal{G}}_{k,i}$, $M_k \in \mathbb{R}^{(|\mathcal{V}_k| - |\mathcal{V}_{k+1}|) \times |\mathcal{V}_{k+1}|}$ is an element-wise non-negative matrix, and $L_{k+1} \in \mathbb{R}^{|\mathcal{V}_{k+1}| \times |\mathcal{V}_{k+1}|}$ is the Laplacian matrix corresponding to the graph $\mathcal{G}_{k+1}$. The matrices $D_k \in \mathbb{R}^{|\mathcal{V}_k| \times |\mathcal{V}_k|}$ are non-negative diagonal matrices defined recursively as $D_{k+1} = D_k^* + \operatorname{diag}(M_k \cdot 1_{|\mathcal{V}_{k+1}|}) + L_{k+1} \in \mathbb{R}^{|\mathcal{V}_{k+1}| \times |\mathcal{V}_{k+1}|}$, $k = 0, \dots, k^* - 1$.

Furthermore, since all vertices of $\mathcal{G}_{k^*}$ belong to an iSCC by construction, we can assume without loss of generality that

$$D_{k^*} + L_{k^*} = \begin{pmatrix} D_{k^*,1} + \tilde{L}_{k^*,1} & & 0 \\ & \ddots & \\ 0 & & D_{k^*,c_{k^*}} + \tilde{L}_{k^*,c_{k^*}} \end{pmatrix}.$$

In Theorem 2.12, we showed that any vertex not belonging to an iSCC of some graph is descendant of vertices that belong to an iSCC. Since the vertices of $\mathcal{G}_{k+1}$ do not belong to an iSCC of $\mathcal{G}_k$ for $k = 0, \dots, k^* - 1$ and the vertices of $\mathcal{G}_0$ do not belong to an iSCC of $\mathcal{G}(t)$, this implies that $D_{k,i} \neq 0$, $k = 0, \dots, k^*$, $i \in \mathbb{N}_{c_k}$, thus $D_{k,i} + \tilde{L}_{k,i}$ is irreducibly diagonally dominant, and therefore by Lemma A.8 invertible. Consequently $D_0 + L_0 = \operatorname{diag}(M_0 \cdot 1_{N_0}) + L_0 \in \mathbb{R}^{N_0 \times N_0}$ is invertible. This completes the proof. □

Appendix B

Asymptotic Properties of Ordinary Differential Equations

B.1 Linear Systems with Exponentially Decaying Inputs

At several places in this thesis, we use the fact that, under certain conditions, solutions of an asymptotically autonomous linear system converge to particular solutions of the corresponding autonomous linear system. More precisely, we consider linear systems of the form

$$\dot{x}(t) = Ax(t) + u(t) \tag{B.1}$$

with state vector $x(t) \in \mathbb{R}^n$ and input $u(\cdot) : \mathbb{R} \to \mathbb{R}^n$, vanishing for $t \to \infty$. We are interested in the asymptotic behavior of solutions $x(t)$ of (B.1) and how these solutions relate to solutions of (B.1) in case $u(t) \equiv 0$. Frequently, the answer to this question is taken as granted although most standard books on linear systems theory lack its proof. If system (B.1) is asymptotically stable, the answer is a particular case of the theory of input to state stability (see Khalil, 2002). For the sake of completeness, we give the answer, for arbitrary systems (B.1) in the case when the input $u(t)$ decays exponentially fast, in the following Lemma:

Lemma B.1. *Let $x(t)$ be any solution of the non-autonomous linear system* (B.1). *Let $M_1, \mu_1 \in \mathbb{R}_+$ and assume the input $u(t)$ satisfies*

$$\|u(t)\| \leq M_1 e^{-\mu_1 t} \tag{B.2}$$

Then there exists some $\tilde{x} \in \mathbb{R}^n$ and constants $M_2, \mu_2 \in \mathbb{R}_+$ such that

$$\|x(t) - e^{At}\tilde{x}\| \leq M_2 e^{-\mu_2 t} \tag{B.3}$$

for all $t \geq 0$.

Proof. Without loss of generality, we assume that $A \in \mathbb{R}^{n\times n}$ is given in Jordan canonical form

$$A = \begin{pmatrix} J_{m_1}(\lambda_1) & & 0 \\ & \ddots & \\ 0 & & J_{m_r}(\lambda_r) \end{pmatrix}$$

with Jordan blocks $J_{m_k}(\lambda_k) \in \mathbb{R}^{m_k \times m_k}$, $k \in \mathbb{N}_r$ for some $r \in \mathbb{N}_n$ with $\sum_{k=1}^{r} m_k = n$. There exist an integer r^* with $0 \leq r^* \leq r$ such that $\lambda_k \in \mathbb{C}_-$ for all $k \in \mathbb{N}_{r^*}$ and $\lambda_k \in \overline{\mathbb{C}}_+$

for all $k \in \mathbb{N}_r \setminus \mathbb{N}_{r^*}$, i.e., the first r^* Jordan blocks have eigenvalues with negative real parts and the remaining Jordan blocks have eigenvalues with non-negative real parts. We consider a partitioning of the the state vector $x(t)$, the input vector $u(t)$, and the vector $\tilde{x}$ corresponding to the Jordan blocks as $x_k(t) \in \mathbb{R}^{m_k}$, $u_k(t) \in \mathbb{R}^{m_k}$, and $\tilde{x}_k \in \mathbb{R}^{m_k}$, $k \in \mathbb{N}_r$.

The solutions $x_k(t)$, $k \in \mathbb{N}_r$ are independently given as

$$x_k(t) = e^{J_{m_k}(\lambda_k)t} x_k(0) + \int_0^t e^{J_{m_k}(\lambda_k)(t-\tau)} u_k(\tau) \mathrm{d}\tau, \quad k \in \mathbb{N}_r.$$

In the sequel, we will make use of upper norm bounds on the matrix exponential $e^{J_m(\lambda)} \in \mathbb{R}^{m \times m}$ of a Jordan block $J_m(\lambda) \in \mathbb{R}^{m \times m}$ for some $m \in \mathbb{N}$ and $\lambda \in \mathbb{C}$. Let $\nu \in \mathbb{R}$ with $\nu > \mathfrak{Re}(\lambda)$. There exists a constant $\alpha \in \mathbb{R}_+$ and a real polynomial $p(t)$ of degree $m-1$ in t satisfying $p(t) \geq 0$ for $t \geq 0$, such that

$$\|e^{J_m(\lambda)}\| \leq p(t) e^{\mathfrak{Re}(\lambda)t} \leq \alpha e^{\nu t}, \quad t \geq 0.$$

These bounds are consequences of properties of the matrix exponential and the special form of a Jordan block. See Horn and Johnson (1991) for details.

We first consider the case that $\lambda_k \in \mathbb{C}_-$. Let $\tilde{\nu}_k \in \mathbb{R}_+$ with $\tilde{\nu}_k < -\mathfrak{Re}(\lambda_k)$. Then there exists a constant $\tilde{\alpha}_k \in \mathbb{R}_+$ such that $\|e^{J_{m_k}(\lambda_k)t}\| \leq \tilde{\alpha}_k e^{-\tilde{\nu}_k t}$ for all $t \geq 0$. Consequently, using (B.2),

$$\begin{aligned}
\|x_k(t)\| &\leq \tilde{\alpha}_k e^{-\tilde{\nu}_k t} \|x_k(0)\| + \int_0^t \tilde{\alpha}_k e^{-\tilde{\nu}_k(t-\tau)} M_1 e^{-\mu_1 \tau} \mathrm{d}\tau \\
&\leq \tilde{\alpha}_k e^{-\min(\tilde{\nu}_k,\mu_1)t} \|x_k(0)\| + \int_0^t \tilde{\alpha}_k M_1 e^{-\min(\tilde{\nu}_k,\mu_1)t} \mathrm{d}\tau \\
&= e^{-\min(\tilde{\nu}_k,\mu_1)t} (\tilde{\alpha}_k \|x_k(0)\| + \tilde{\alpha}_k M_1 t).
\end{aligned}$$

Hence, for any $k \in \mathbb{N}_{r^*}$ and any $\nu_k \in \mathbb{R}_+$ with $0 < \nu_k < \min(\tilde{\nu}_k, \mu_1)$, there exists a constant $\alpha_k \in \mathbb{R}_+$ such that

$$\|x_k(t)\| \leq \alpha_k e^{-\nu_k t}, \quad t \geq 0. \tag{B.4}$$

That is, $(x_k(t) - e^{J_{m_k}(\lambda)t} \tilde{x}_k) \to 0$ as $t \to \infty$ with $\tilde{x}_k = 0$.

Next, we consider the case that $\lambda_k \in \overline{\mathbb{C}}_+$. We rewrite the solution as

$$x_k(t) = e^{J_{m_k}(\lambda_k)t} \left(x_k(0) + \int_0^t e^{-J_{m_k}(\lambda_k)\tau} u_k(\tau) \mathrm{d}\tau \right).$$

There exists a real polynomial $p_k(t)$ of degree $m_k - 1$ in t satisfying $p_k(t) \geq 0$ for $t \geq 0$ such that $\|e^{-J_{m_k}(\lambda_k)t}\| \leq p_k(t) e^{-\mathfrak{Re}(\lambda_k)t}$ for $t \geq 0$. Using (B.2), this implies that the integrand in the above solution is norm-bounded as

$$\left\| e^{-J_{m_k}(\lambda_k)t} u_k(t) \right\| \leq p_k(t) M_1 e^{-(\mathfrak{Re}(\lambda_k)+\mu_1)t}.$$

Let $\overline{\nu}_k \in \mathbb{R}_+$ with $\mathfrak{Re}(\lambda_k) < \overline{\nu}_k < \mathfrak{Re}(\lambda_k) + \mu_1$ (we choose $\overline{\nu}_k < \mathfrak{Re}(\lambda_k) + \mu_1$ in order to obtain a bound for the integrand which does not depend on a polynomial; the reason for the lower bound $\overline{\nu}_k > \mathfrak{Re}(\lambda_k)$ will become apparent later). Then there exist constants $\overline{\alpha}_k \in \mathbb{R}_+$ such that $p_k(t) M_1 e^{-(\mathfrak{Re}(\lambda_k)+\mu_1)t} \leq \overline{\alpha}_k e^{-\overline{\nu}_k t}$. Hence, the integrands in the above solutions $x_k(t)$ are norm-bounded by exponentially decaying functions. This implies (see Apostol, 1974, Section 10.13) that there exist vectors $\xi_k \in \mathbb{R}^{m_k}$ such that

$$\int_0^\infty e^{-J_{m_k}(\lambda_k)\tau} u_k(\tau) \mathrm{d}\tau = \xi_k.$$

Moreover, since the integrals exist, we have

$$\left\| \int_0^t e^{-J_{m_k}(\lambda_k)\tau} u_k(\tau) \mathrm{d}\tau - \xi_k \right\| = \left\| \int_t^\infty e^{-J_{m_k}(\lambda_k)\tau} u_k(\tau) \mathrm{d}\tau \right\| \leq \int_t^\infty \overline{\alpha}_k e^{-\overline{\nu}_k \tau} \mathrm{d}\tau = \frac{\overline{\alpha}_k}{\overline{\nu}_k} e^{-\overline{\nu}_k t}.$$

Let $\hat{\nu}_k \in \mathbb{R}_+$ with $\mathfrak{Re}(\lambda_k) < \hat{\nu}_k < \overline{\nu}_k$ (such a constant $\hat{\nu}_k$ exists because we chose $\overline{\nu}_k > \mathfrak{Re}(\lambda_k)$ before). Then there exist constants $\hat{\alpha}_k$ such that $\|e^{J_{m_k}(\lambda_k)t}\| \leq \hat{\alpha}_k e^{\hat{\nu}_k t}$ for all $t \geq 0$. Finally, we combine the bounds on $\left\| \int_0^t e^{-J_{m_k}(\lambda_k)\tau} u_k(\tau) \mathrm{d}\tau - \xi_k \right\|$ and $\|e^{J_{m_k}(\lambda_k)t}\|$: With $\alpha_k = \frac{\hat{\alpha}_k \overline{\alpha}_k}{\overline{\nu}_k}$ and $\nu_k = \overline{\nu}_k - \hat{\nu}_k > 0$, $k \in \mathbb{N}_r \setminus \mathbb{N}_{r^*}$ we obtain

$$\left\| x_k(t) - e^{J_{m_k}(\lambda_k)t}(x_k(0) + \xi_k) \right\| \leq \alpha_k e^{-\nu_k t}, \quad t \geq 0. \tag{B.5}$$

That is $(x_k(t) - e^{J_{m_k}(\lambda_k)t}\tilde{x}_k) \to 0$ as $t \to \infty$ with $\tilde{x}_k = x_k(0) + \xi_k$.

Combining (B.4) and (B.5), this implies that (B.3) is satisfied with $\tilde{x} = (\tilde{x}_1^T, \ldots, \tilde{x}_r^T)^T$, where $\tilde{x}_k = 0 \in \mathbb{R}^{m_k}$ for $k \in \mathbb{N}_{r^*}$, $\tilde{x}_k = x_k(0) + \xi_k \in \mathbb{R}^{m_k}$ for $k \in \mathbb{N}_r \setminus \mathbb{N}_{r^*}$, $M_2 = \sqrt{r} \max_{k \in \mathbb{N}_r} \alpha_k$, and $\mu_2 = \min_{k \in \mathbb{N}_r} \nu_k$. □

B.2 Limit Sets

If one is interested in the asymptotic (or *steady-state*) behavior of a system modeled by nonlinear ODEs, limit sets prove to be a useful tool. Below, we will give the most basic definitions and properties for autonomous and asymptotically autonomous (see Mischaikow et al., 1995, Strauss and Yorke, 1967) systems. Further details, including detailed proofs, of the notions and results presented below can be found in Hahn (1967), Hale (1988), Hale et al. (2002). The importance of limit sets of sets for the characterization of the steady-state behavior of nonlinear systems is stressed in Byrnes and Isidori (2003), Isidori and Byrnes (2008).

B.2.1 Autonomous Systems

Consider a dynamical system modeled by the set of first order ODEs

$$\dot{x}(t) = f(x(t)) \tag{B.6}$$

with state vector $x(t) \in \mathbb{R}^n$ and time $t \in \mathbb{R}$. Assume $f : \mathbb{R}^n \to \mathbb{R}^n$ is locally Lipschitz, i.e., solutions of (B.6) with initial condition $x(0) = x_0 \in \mathbb{R}^n$ exist in some open interval containing the point $t = 0$ (see Khalil, 2002). We denote the solution of (B.6) passing through $x_0 \in \mathbb{R}^n$ at time $t = 0$ as $x(t; x_0)$.

If the solution $x(t; x_0)$ to (B.6) is defined for all $t \in \overline{\mathbb{R}}_+$, the *positive orbit through* x_0 is defined as $\Gamma^+(x_0) \triangleq \bigcup_{t \in \overline{\mathbb{R}}_+} x(t; x_0)$. Likewise, if $x(t; x_0)$ is defined for all $t \in \mathbb{R}$, the *complete orbit through* x_0 is defined as $\Gamma(x_0) \triangleq \bigcup_{t \in \mathbb{R}} x(t; x_0)$.

The limiting behavior of a single solution $x(t; x_0)$ of (B.6) can be characterized by the *ω-limit set of x_0* (or the *ω-limit set of the positive orbit* $\Gamma^+(x_0)$). If $\Gamma^+(x_0)$ exists for some initial condition $x_0 \in \mathbb{R}^n$, the *ω-limit set of x_0* is defined as

$$\omega(x_0) = \bigcap_{t \geq 0} \mathrm{Cl}\left(\Gamma^+(x(t; x_0))\right).$$

This definition is equivalent to saying that $\hat{x} \in \omega(x_0)$ if and only if there exists a sequence $\{t_k\}$, $k \in \mathbb{N}$ with $t_k \to \infty$ as $k \to \infty$ such that $x(t_k, x_0) \to \hat{x}$ as $k \to \infty$. That is, the points contained in $\omega(x_0)$ are precisely those points to which the solution $x(t; x_0)$ gets arbitrarily close for arbitrarily large times.

The ω-limit set of a point enjoys some nice properties summarized in the following lemma (see Hahn, 1967, Chapter III):

Lemma B.2. *Given system* (B.6) *with initial condition $x_0 \in \mathbb{R}^n$. Suppose $x(t; x_0)$ exists and is bounded for all $t \geq 0$.*

Then $\omega(x_0)$ is a nonempty connected compact set such that $\inf_{y \in \omega(x)} \|x(t; x_0) - y\| \to 0$ *as $t \to \infty$. Furthermore, for any $y \in \omega(x_0)$, $x(t; y)$ is defined for all $t \in \mathbb{R}$ and $\Gamma(y) \subset \omega(x_0)$.*

That is, the ω-limit set of a bounded orbit is guaranteed to exist, to attract the orbit, and to be an invariant set for (B.6).

All the definitions and results stated as properties of points above are readily extended to properties of sets. Given system (B.6) and some set $\mathcal{B} \subset \mathbb{R}^n$. The *positive and complete orbit through* $\mathcal{B}$ is defined as $\Gamma^+(\mathcal{B}) = \bigcup_{x \in \mathcal{B}} \Gamma^+(x) = \bigcup_{t \in \overline{\mathbb{R}}_+} x(t; \mathcal{B})$ and $\Gamma(\mathcal{B}) = \bigcup_{x \in \mathcal{B}} \Gamma(x) = \bigcup_{t \in \mathbb{R}} x(t; \mathcal{B})$ respectively, if the corresponding orbits $\Gamma^+(x)$ and $\Gamma(x)$ are defined for all $x \in \mathcal{B}$.

With the help of the orbit of a set, invariance of a set under (B.6) is conveniently characterized as follows: A set $\mathcal{B} \subset \mathbb{R}^n$ is called *invariant* under (B.6) if $\Gamma^+(\mathcal{B}) = \mathcal{B}$. It is called positively invariant under (B.6) if $\Gamma^+(\mathcal{B}) \subset \mathcal{B}$.

The *ω-limit set of a set* $\mathcal{B} \subset \mathbb{R}^n$, for which $\Gamma^+(\mathcal{B})$ exists, is defined as

$$\omega(\mathcal{B}) = \bigcap_{t \geq 0} \mathrm{Cl}\left(\Gamma^+(x(t; \mathcal{B}))\right).$$

This definition is equivalent to saying that $\hat{x} \in \omega(\mathcal{B})$ if and only if there exists sequences $\{t_k\}$ and $\{x_k\}$, $k \in \mathbb{N}$ with $t_k \to \infty$ as $k \to \infty$ and $x_k \in \mathcal{B}$, $k \in \mathbb{N}$ such that $x(t_k; x_k) \to \hat{x}$ as $k \to \infty$. That is, the points contained in $\omega(\mathcal{B})$ are precisely those points to which the family of solutions $x(t; \mathcal{B})$ gets arbitrarily close for arbitrarily large times.

Note that $\omega(\mathcal{B}) \supset \bigcup_{x \in \mathcal{B}} \omega(x)$ by definition while equality does not hold in general (see Hale, 1988, Hale et al., 2002). The generalization of Lemma B.2 to limit sets of sets can be found in Hale et al. (2002, Lemma 2.0.1) (in a much more general version) and is repeated below.

Lemma B.3. *Given system* (B.6) *and some set $\mathcal{B} \subset \mathbb{R}^n$. Suppose $\mathcal{B}$ is nonempty and $\Gamma^+(\mathcal{B})$ exists and is bounded (which implies boundedness of $\mathcal{B}$).*

Then $\omega(B)$ is a nonempty compact set such that

$$\lim_{t \to \infty} \sup_{\hat{x} \in x(t; \mathcal{B})} \inf_{\check{x} \in \omega(\mathcal{B})} \|\hat{x} - \check{x}\| = 0. \tag{B.7}$$

Furthermore, $\Gamma^+(\omega(\mathcal{B}))$ is defined and $\Gamma^+(\omega(\mathcal{B})) = \omega(\mathcal{B})$. If $\mathcal{B}$ is connected, so is $\omega(\mathcal{B})$.

Property (B.7) means that all points of $\mathcal{B}$ are uniformly attracted by $\omega(\mathcal{B})$. We say that $\omega(\mathcal{B})$ attracts the set $\mathcal{B}$. Lemma B.2 is obviously a special case of Lemma B.3. It is recovered by setting $\mathcal{B} = \{x_0\}$.

We conclude the section about limit sets for autonomous systems with an important result adapted from Byrnes and Isidori (2003, Lemma 4.1).

Lemma B.4. *Given system* (B.6) *and some set* $\mathcal{B} \subset \mathbb{R}^n$. *Suppose* $\mathcal{B}$ *is nonempty and* $\Gamma^+(\mathcal{B})$ *exists and is bounded. Let* $\mathcal{K} \subset \mathbb{R}^n$ *be some closed set.*

Then

$$\lim_{t\to\infty} \sup_{\hat{x}\in x(t;\mathcal{B})} \inf_{\check{x}\in\mathcal{K}} \|\hat{x} - \check{x}\| = 0 \tag{B.8}$$

if and only if $\omega(\mathcal{B}) \subset \mathcal{K}$.

Proof. Sufficiency is immediate from Lemma B.3.

Now suppose (B.8) holds and $\hat{x} \in \omega(\mathcal{B})$. Then there exist sequences $\{t_k\}$, $\{x_k\}$, $k \in \mathbb{N}$ with $t_k \to \infty$ as $k \to \infty$ and $x_k \in \mathcal{B}$, $k \in \mathbb{N}$ with the property that for any $\varepsilon > 0$, there exists $\tilde{k}$ such that $\|x(t_k, x_k) - \hat{x}\| \leq \varepsilon$ for all $k \geq \tilde{k}$. On the other hand (B.8) implies that for any $\varepsilon > 0$, there exists $\bar{t}$ such that $\inf_{\check{x}\in\mathcal{K}} \|x(t;x_0) - \check{x}\| \leq \varepsilon$ for any $x_0 \in \mathcal{B}$ and all $t \geq \bar{t}$. Since $t_k \to \infty$ as $k \to \infty$, there exists $\bar{k} \geq \tilde{k}$ such that $t_k \geq \bar{t}$ for all $k \geq \bar{k}$. Then, using the triangle inequality, it follows that $\inf_{\check{x}\in\mathcal{K}} \|\hat{x} - \check{x}\| \leq 2\varepsilon$, i.e., since $\varepsilon > 0$ is arbitrary, $\inf_{\check{x}\in\mathcal{K}} \|\hat{x} - \check{x}\| = 0$. Thus, since $\mathcal{K}$ is closed, $\hat{x} \in \mathcal{K}$. Since $\hat{x} \in \omega(\mathcal{B})$ is arbitrary, $\omega(\mathcal{B}) \subset \mathcal{K}$ follows. □

Note that (B.8) implies uniform convergence of all solutions starting in $\mathcal{B}$ to the set $\mathcal{K}$, i.e., given an ε-neighborhood of $\mathcal{K}$, the time for any solution starting in $\mathcal{B}$ to enter the neighborhood does not depend on the initial condition. If one drops this requirement and replaces (B.8) by the weaker notion that $\inf_{\check{x}\in\mathcal{K}} \|x(t;x_0) - \check{x}\| \to 0$ as $t \to \infty$ for all $x_0 \in \mathcal{B}$, a necessary and sufficient condition is $\bigcup_{x_0\in\mathcal{B}} \omega(x_0) \subset \mathcal{K}$. Since in general $\bigcup_{x_0\in\mathcal{B}} \omega(x_0)$ is a proper subset of $\omega(\mathcal{B})$, the property that $\inf_{\check{x}\in\mathcal{K}} \|x(t;x_0) - \check{x}\| \to 0$ as $t \to \infty$ for all $x_0 \in \mathcal{B}$ does not imply $\omega(\mathcal{B}) \subset \mathcal{K}$ in general.

The difference between the limit set $\omega(\mathcal{B})$ and the union of the limit sets $\bigcup_{x\in\mathcal{B}} \omega(x)$ is illustrated in Figure B.1 when (B.6) is the Van der Pol oscillator. Figure B.1(a) shows the case when $\mathcal{B}$ is a compact set containing the limit cycle together with the set enclosed by the limit cycle. The set $\bigcup_{x\in\mathcal{B}} \omega(x)$ is depicted in bold black. It consists of the limit cycle and the unstable equilibrium. The limit set $\omega(\mathcal{B})$ is depicted in dark grey. It consists of the limit cycle and the entire set enclosed by the limit cycle. This fact is verified as follows: Consider any point x_0 in the set enclosed by the limit cycle. Then the solution $x(t;x_0)$ is defined for all $t \in \mathbb{R}$ and remains in that set. Let $\{t_k\}$, $k \in \mathbb{N}$ be any sequence such that $t_k \to \infty$ as $k \to \infty$ and define $x_k = x(-t_k;x_0) \in \mathcal{B}$, $k \in \mathbb{N}$. Then $x(t_k, x_k) = x_0$ for all $k \in \mathbb{N}$ which proves that $x_0 \in \omega(\mathcal{B})$. Note that the points x_k are contained in $\omega(\mathcal{B})$ but not in $\bigcup_{x\in\mathcal{B}} \omega(x)$. Since for any time $\bar{t} \in \mathbb{R}_+$, there exists $\bar{k} \in \mathbb{N}$ such that $t_k > \bar{t}$ for any $k \geq \bar{k}$, i.e., the time needed to reach x_0 starting from x_k is at least $\bar{t}$ for all $k \geq \bar{k}$. This shows that convergence to $\bigcup_{x\in\mathcal{B}} \omega(x)$ cannot be uniform in the initial condition $x \in \mathcal{B}$ despite $\mathcal{B}$ being compact. Such situations can be excluded by appropriately modifying the set $\mathcal{B}$. Figure B.1(b) depicts the situation when the unstable equilibrium together with an open neighborhood is excluded from $\mathcal{B}$, i.e., $\mathcal{B}$ contains no unstable invariant set. In that case, the only solutions contained in $\mathcal{B}$ for all times $t \in \mathbb{R}$ are those which are contained in $\bigcup_{x\in\mathcal{B}} \omega(x)$, thus $\omega(\mathcal{B}) = \bigcup_{x\in\mathcal{B}} \omega(x)$.

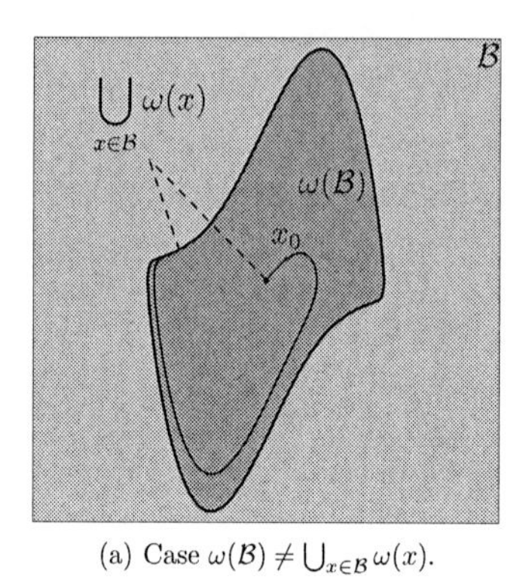

(a) Case $\omega(\mathcal{B}) \neq \bigcup_{x\in\mathcal{B}} \omega(x)$.

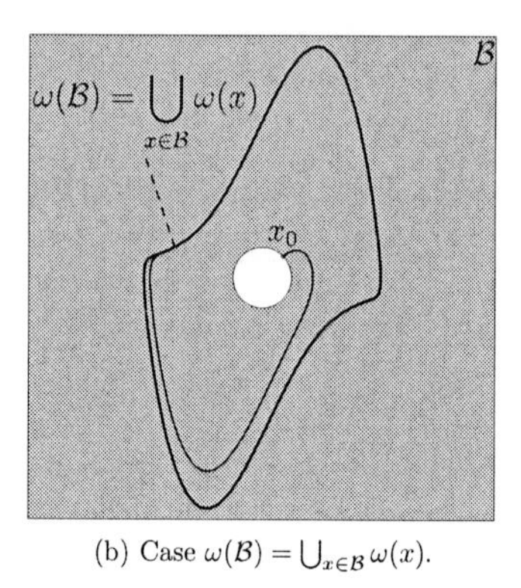

(b) Case $\omega(\mathcal{B}) = \bigcup_{x\in\mathcal{B}} \omega(x)$.

Figure B.1: Limit sets for the Van der Pol oscillator.

B.2.2 Asymptotically Autonomous Systems

Throughout this thesis, we often investigate systems which are coupled over time-varying communication topologies. This leads to systems of ODEs that explicitly depend on time, i.e., to non-autonomous ODEs of the form

$$\dot{x}(t) = f(t, x(t)) \tag{B.9}$$

with state vector $x(t) \in \mathbb{R}^n$ and time $t \in \mathbb{R}$. We assume that $f : \mathbb{R} \times \mathbb{R}^n \to \mathbb{R}^n$ is piecewise continuous in t and locally Lipschitz in x to guarantee existence and uniqueness of solutions as in the case of autonomous systems considered before. Since solutions depend on initial conditions and initial time for systems of the type (B.9), we denote solutions of (B.9) passing through $x_0 \in \mathbb{R}^n$ at time $t = t_0 \in \mathbb{R}$ as $x(t; (t_0, x_0))$.

We can define orbits and limit sets as in the autonomous case. We skip the definition for single solutions, as they are obtained as special cases of orbits and limit sets of sets. Consider some extended set $\mathcal{B}_e \subset \mathbb{R} \times \mathbb{R}^n$ of initial times and initial states. To simplify notation, we define the set $X(t; \mathcal{B}_e) \triangleq \bigcup_{(t_0,x_0)\in\mathcal{B}_e} x(t_0 + t; (t_0, x_0))$. If $x(t_0 + t; (t_0, x_0))$ exists for all $t \in \overline{\mathbb{R}}_+$ and all $(t_0, x_0) \in \mathcal{B}_e$, the positive orbit through $\mathcal{B}_e$ is defined as $\Gamma^+(\mathcal{B}_e) \triangleq \bigcup_{t\in\overline{\mathbb{R}}_+} X(t; \mathcal{B}_e)$. If $x(t_0 + t; (t_0, x_0))$ exists for all $t \in \mathbb{R}$ and all $(t_0, x_0) \in \mathcal{B}_e$, the complete orbit through $\mathcal{B}_e$ is defined as $\Gamma(\mathcal{B}_e) \triangleq \bigcup_{t\in\mathbb{R}} X(t; \mathcal{B}_e)$. The ω-limit set of $\mathcal{B}_e$ is defined by

$$\omega(\mathcal{B}_e) \triangleq \bigcap_{t\in\overline{\mathbb{R}}_+} \operatorname{Cl} \bigcup_{\tau\geq t} X(\tau; \mathcal{B}_e).$$

Note that $\Gamma^+(\mathcal{B}_e) \subset \mathbb{R}^n$, $\Gamma(\mathcal{B}_e) \subset \mathbb{R}^n$, and $\omega(\mathcal{B}_e) \subset \mathbb{R}^n$ are subsets of the state space $\mathbb{R}^n$ while the argument $\mathcal{B}_e \subset \mathbb{R} \times \mathbb{R}^n$ is a set of initial states and *initial times*. Thus, in particular, we cannot expect orbits or limit sets to be invariant in general, i.e., for general non-autonomous systems, limit sets are significantly less useful than in the case of autonomous systems considered in Section B.2.1.

However, some of the results can be generalized to the case of non-autonomous systems. For instance, Lemma B.3 can partly be generalized to non-autonomous systems as follows:

Lemma B.5. *Given system* (B.9) *and some set* $\mathcal{B}_e \subset \mathbb{R} \times \mathbb{R}^n$*. Suppose* $\mathcal{B}_e$ *is nonempty and* $\Gamma^+(\mathcal{B}_e)$ *exists and is bounded.*

Then $\omega(\mathcal{B}_e)$ *is a nonempty compact set such that*

$$\lim_{t\to\infty} \sup_{\hat{x}\in X(t;\mathcal{B}_e)} \inf_{\check{x}\in\omega(\mathcal{B}_e)} \|\hat{x} - \check{x}\| = 0. \tag{B.10}$$

Proof. Suppose $\Gamma^+(\mathcal{B}_e)$ exists and is bounded, then $\mathrm{Cl}\bigcup_{\tau\geq t} X(\tau;\mathcal{B}_e)$ is nonempty and compact for any $t \geq 0$ and, by definition of limit-sets, $\omega(\mathcal{B}_e)$ is nonempty and compact.

We now prove (B.10) by contradiction. Suppose (B.10) does not hold. Then there exist $\varepsilon \in \mathbb{R}_+$ and sequences $\{t_k\}$, $\{(\tau_k, x_k)\}$, $k \in \mathbb{N}$ with $t_k \geq 0$, $t_k \to \infty$ as $k \to \infty$, and $(\tau_k, x_k) \in \mathcal{B}_e$, $k \in \mathbb{N}$ such that $\inf_{\check{x}\in\omega(\mathcal{B}_e)} \|x(\tau_k + t_k; (\tau_k, x_k)) - \check{x}\| > \varepsilon$ for all $k \in \mathbb{N}$. Since $x(\tau_k+t_k; (\tau_k, x_k)) \in \mathrm{Cl}\,\Gamma^+(\mathcal{B}_e)$, which is compact, the sequence $\{x(\tau_k+t_k; (\tau_k, x_k))\}$, $k \in \mathbb{N}$ has a convergent subsequence. Since the limit of this sequence must belong to $\omega(\mathcal{B}_e)$, we obtain a contradiction. Thus (B.10) holds. □

With this, it is also easy to generalize Lemma B.4, which is done below.

Lemma B.6. *Given system* (B.9) *and some set* $\mathcal{B}_e \subset \mathbb{R} \times \mathbb{R}^n$*. Suppose* $\mathcal{B}_e$ *is nonempty and* $\Gamma^+(\mathcal{B}_e)$ *exists and is bounded. Let* $\mathcal{K} \subset \mathbb{R}^n$ *be some closed set.*

Then

$$\lim_{t\to\infty} \sup_{\hat{x}\in X(t;\mathcal{B}_e)} \inf_{\check{x}\in\mathcal{K}} \|\hat{x} - \check{x}\| = 0 \tag{B.11}$$

if and only if $\omega(\mathcal{B}_e) \subset \mathcal{K}$.

Proof. Sufficiency is immediate from Lemma B.5.

To prove necessity, suppose (B.11) holds and $\hat{x} \in \omega(\mathcal{B}_e)$. Then there exist sequences $\{t_k\}$, $\{\tau_k, x_k\}$, $k \in \mathbb{N}$ with $t_k \to \infty$ as $k \to \infty$ and $(\tau_k, x_k) \in \mathcal{B}_e$, $k \in \mathbb{N}$ with the property that for any $\varepsilon > 0$, there exists $\tilde{k}$ such that $\|x(\tau_k + t_k; (\tau_k, x_k)) - \hat{x}\| \leq \varepsilon$ for all $k \geq \tilde{k}$. On the other hand (B.11) implies that for any $\varepsilon > 0$, there exists $\bar{t}$ such that $\inf_{\check{x}\in\mathcal{K}} \|x(t_0 + t; (t_0, x_0)) - \check{x}\| \leq \varepsilon$ for any $(t_0, x_0) \in \mathcal{B}_e$ and all $t \geq \bar{t}$. Since $t_k \to \infty$ as $k \to \infty$, there exists $\bar{k} \geq \tilde{k}$ such that $t_k \geq \bar{t}$ for all $k \geq \bar{k}$. Then, using the triangle inequality, it follows that $\inf_{\check{x}\in\mathcal{K}} \|\hat{x} - \check{x}\| \leq 2\varepsilon$, i.e., since $\varepsilon > 0$ is arbitrary, $\inf_{\check{x}\in\mathcal{K}} \|\hat{x} - \check{x}\| = 0$. Thus, since $\mathcal{K}$ is closed, $\hat{x} \in \mathcal{K}$. Since $\hat{x} \in \omega(\mathcal{B}_e)$ is arbitrary, $\omega(\mathcal{B}_e) \subset \mathcal{K}$ follows. □

To obtain generalizations of other results for limit sets, in particular results related to invariance properties, we need to consider special cases of non-autonomous systems. One such special case in which limit sets for non-autonomous systems inherit much of the useful properties of limit sets for autonomous systems, is the case of asymptotically autonomous systems. Consider an autonomous system modeled as the set of ODEs

$$\dot{\overline{x}}(t) = \overline{f}(\overline{x}(t)), \tag{B.12}$$

with state vector $\overline{x}(t) \in \mathbb{R}^n$. System (B.9) is called asymptotically autonomous with limit system (B.12) if

$$\lim_{t\to\infty} \|f(t,x) - \overline{f}(x)\| = 0 \tag{B.13}$$

uniformly on every compact subset of $\mathbb{R}^n$ (see Mischaikow et al., 1995). Intuitively, since ω-limit sets describe the asymptotic behavior of a system and system (B.9) is asymptotically equal to an autonomous system (B.12), what one would hope for is that ω-limit sets are a meaningful notion for the limit system (B.12). And that is exactly what one obtains.

We skip all the details and just present the basic idea and the result below. For a thorough presentation, the reader is referred to Mischaikow et al. (1995), Strauss and Yorke (1967). We define $z(t) = (s(t), x(t)) \subset [-\infty, \infty] \times \mathbb{R}^n$ where $[-\infty, \infty]$ is the extended real line, i.e., a compact set homeomorphic to the unit interval $[1, 1] \subset \mathbb{R}$. Then

$$\dot{z}(t) = \begin{cases} (1, f(s(t), x(t))) & \text{if } s(t) \neq \infty, \\ (0, \overline{f}(x(t))) & \text{if } s(t) = \infty, \end{cases} \tag{B.14}$$

which is an autonomous system. The limit sets for system (B.14) are all contained in $\{\infty\} \times \mathbb{R}^n \subset [-\infty, \infty] \times \mathbb{R}^n$. In fact, given a set of initial conditions $\mathcal{B}_e \subset [-\infty, \infty] \times \mathbb{R}^n$, the limit set for (B.14) is $\{\infty\} \times \omega(\mathcal{B}_e)$ where $\omega(\mathcal{B}_e)$ is the limit set for (B.9) defined before. Thus, we can extend Lemma B.3 to asymptotically autonomous systems as follows:

Lemma B.7. *Given system* (B.9) *with limit system* (B.12). *Assume* (B.13) *is satisfied uniformly in any compact subset of* $\mathbb{R}^n$. *Given a set* $\mathcal{B}_e \subset \mathbb{R} \times \mathbb{R}^n$. *Suppose* $\mathcal{B}_e$ *is nonempty and* $\Gamma^+(\mathcal{B}_e)$ *exists and is bounded.*

Then $\omega(\mathcal{B}_e)$ *is a nonempty compact set such that*

$$\lim_{t\to\infty} \sup_{\hat{x}\in X(t;\mathcal{B}_e)} \inf_{\check{x}\in\omega(\mathcal{B}_e)} \|\hat{x} - \check{x}\| = 0. \tag{B.15}$$

Furthermore, $\omega(\mathcal{B}_e)$ *is invariant for the limit system* (B.12), *i.e.,* $\overline{x}(t; \omega(\mathcal{B}_e))$ *exists and satisfies* $\overline{x}(t; \omega(\mathcal{B}_e)) = \omega(\mathcal{B}_e)$ *for all times* $t \in \mathbb{R}$. *If* $\mathcal{B}_e$ *is connected, so is* $\omega(\mathcal{B}_e)$.

Note that boundedness of $\Gamma^+(\mathcal{B}_e)$ does not imply boundedness of $\mathcal{B}_e$ in $\mathbb{R}\times\mathbb{R}^n$. However, if $P : \mathbb{R}\times\mathbb{R}^n$ is the projection defined as $(t_0, x_0) \mapsto x_0$, then boundedness of $\Gamma^+(\mathcal{B}_e)$ implies boundedness of $P\mathcal{B}_e = X(0, \mathcal{B}_e)$ in $\mathbb{R}^n$.

Appendix C

Geometric Concepts from Linear Control Theory

In what follows, we investigate properties of linear maps $A : \mathcal{X} \to \mathcal{Y}$ where $\mathcal{X} \simeq \mathbb{R}^n$ and $\mathcal{Y} \simeq \mathbb{R}^m$ are linear vector spaces. For given bases of $\mathcal{X}$ and $\mathcal{Y}$, A can be represented by some matrix $\mathrm{Mat}A$ (see e.g. Wonham, 1985). We will not distinguish sharply between maps A and its matrix representation $\mathrm{Mat}A$ and write simply $A = \mathrm{Mat}A$ if the bases that are used are clear from the context.

C.1 Factor Spaces

In Chapter 4 and Section C.2.1 below, we need the concept of factor spaces of real vector spaces. We give some basic definitions and ideas here. For details, the reader is referred to Wonham (1985) and the references given there.

Consider a subspace $\mathcal{X} \subseteq \mathbb{R}^n$. We call two vectors $x, y \in \mathbb{R}^n$ equivalent mod$\mathcal{X}$ if $(x - y) \in \mathcal{X}$. The factor space $\mathbb{R}^n/\mathcal{X}$ is defined as the set of all equivalence classes

$$x + \mathcal{X} \triangleq \{y : y \in \mathbb{R}^n, y - x \in \mathcal{X}\}, \quad x \in \mathbb{R}^n.$$

The factor space $\mathbb{R}^n/\mathcal{X}$ is a real vector space of dimension $n - \dim(\mathcal{X})$ with

$$\begin{aligned} (x + \mathcal{X}) + (y + \mathcal{X}) &= (x + y) + \mathcal{X}, && x, y \in \mathbb{R}^n, \\ c(x + \mathcal{X}) &= (cx) + \mathcal{X}, && x \in \mathbb{R}^n,\ c \in \mathbb{R}. \end{aligned}$$

Note that $\mathbb{R}^n/\mathcal{X}$ is not a subspace of $\mathbb{R}^n$.

Given a map $A : \mathbb{R}^n \to \mathbb{R}^n$ and a subspace $\mathcal{X} \subseteq \mathbb{R}^n$ invariant under A, i.e., $A\mathcal{X} \subseteq \mathcal{X}$. We denote by $A/\mathcal{X} : \mathbb{R}^n/\mathcal{X} \to \mathbb{R}^n/\mathcal{X}$ the map $x + \mathcal{X} \mapsto (Ax) + \mathcal{X}$.

Let $P : \mathbb{R}^n \to \mathbb{R}^n/\mathcal{X}$ be the canonical projection defined as $x \mapsto x + \mathcal{X}$. Then the map $A/\mathcal{X}$ is uniquely defined by $(A/\mathcal{X})P = PA$. The situation is illustrated in the commutative diagram in Figure C.1.

C.2 Observers, Observability, and Detectability

C.2.1 Observers and Conditionally Invariant Subspaces

Given two subspaces $\mathcal{X}_1, \mathcal{X}_2 \subseteq \mathbb{R}^n$, we use the standard convention that $\mathcal{X}_1 + \mathcal{X}_2 \subseteq \mathbb{R}^n$ denotes the subspace $\{x_1 + x_2 \in \mathbb{R}^n : x_1 \in \mathcal{X}_1, x_2 \in \mathcal{X}_2\} \subseteq \mathbb{R}^n$. Below, we need the

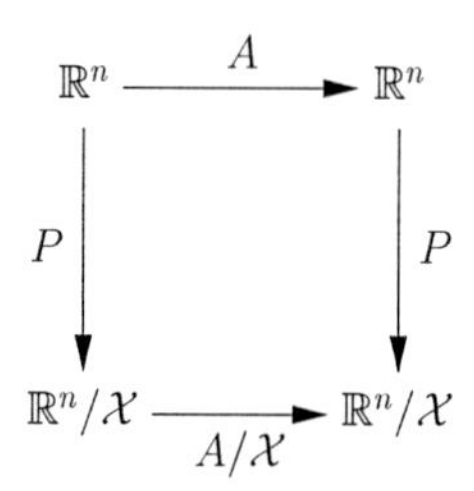

Figure C.1: Commutative diagram for a map $A : \mathbb{R}^n \to \mathbb{R}^n$ and the induced map $A/\mathcal{X} : \mathbb{R}^n/\mathcal{X} \to \mathbb{R}^n/\mathcal{X}$ for an A-invariant subspace $\mathcal{X} \subseteq \mathbb{R}^n$ with the canonical projection $P : \mathbb{R}^n \to \mathbb{R}^n/\mathcal{X}$.

following notational conventions: Given a map $A : \mathbb{R}^n \to \mathbb{R}^n$ and a subspace $\mathcal{X} \subseteq \mathbb{R}^n$, we denote as $A\mathcal{X}$ the subspace obtained as the image of $\mathcal{X}$ under A and by $A^{-1}\mathcal{X} \subseteq \mathbb{R}^n$ the subspace $\{x \in \mathbb{R}^n | \exists y \in \mathbb{R}^n : x = Ay\}$. In particular, if A is represented by some matrix and A^+ is the Moore-Penrose generalized inverse of A (see Horn and Johnson, 1985), then $A^{-1}\mathcal{X} = A^+\mathcal{X} + \ker(A)$. We denote by $A|\mathcal{X} : \mathcal{X} \to A\mathcal{X}$ the restriction of the map A to $\mathcal{X}$ defined as $x \mapsto Ax$, $x \in \mathcal{X}$.

Consider an LTI system

$$\dot{x}(t) = Ax(t) + Bu(t) \tag{C.1a}$$
$$y(t) = Cx(t) \tag{C.1b}$$

with state vector $x(t) \in \mathbb{R}^n$, input vector $u(t) \in \mathbb{R}^p$, and output vector $y(t) \in \mathbb{R}^q$. The state space $\mathbb{R}^n$ of system (C.1) possesses some important geometric properties related to the construction of observers. The concepts introduced below can be found in much more detail e.g. in Willems and Commault (1981), Wonham (1985).

We start by formally defining conditionally invariant subspaces, complementary detectability subspaces and complementary observability subspaces.

Definition C.1 (Conditionally Invariant Subspace, Friend). *Given system* (C.1), *a subspace* $\mathcal{S} \subseteq \mathbb{R}^n$ *is called* conditionally invariant, *if there exists a matrix* $J \in \mathbb{R}^{n \times q}$ *such that* $(A + JC)\mathcal{S} \subseteq \mathcal{S}$. *Such a matrix* J *is called a* friend *of* $\mathcal{S}$. *The set of all friends of* $\mathcal{S}$ *is denoted as* $\boldsymbol{J}(\mathcal{S}) \subseteq \mathbb{R}^{n \times q}$. *The set of all conditionally invariant subspaces is denoted as* $\mathbb{S}$.

Definition C.2 (Complementary Detectability Subspace). *Let* $\mathbb{C} = \mathbb{C}_g \bigcup \mathbb{C}_b$, $\mathbb{C}_g \bigcap \mathbb{C}_b = \emptyset$ *be a symmetric partitioning of the complex plane (where* $\mathbb{C}_g$ *is the 'good', e.g. stable, part and* $\mathbb{C}_b$ *is the 'bad', e.g. unstable, part; symmetric means* $x \in \mathbb{C}_g$ *if and only if* $\overline{x} \in \mathbb{C}_g$*). Given system* (C.1), *a subspace* $\mathcal{S}_g \subseteq \mathbb{R}^n$ *is a* complementary detectability subspace *with respect to* $\mathbb{C}_g$ *if* $\mathcal{S}_g \in \mathbb{S}$ *and there exists a matrix* $J \in \boldsymbol{J}(\mathcal{S}_g)$ *such that* $\sigma((A+JC)/\mathcal{S}_g) \subseteq \mathbb{C}_g$. *The set of all complementary detectability subspaces is denoted as* $\mathbb{S}_g$.

Definition C.3 (Complementary Observability Subspace). *Given system* (C.1), *a subspace* $\mathcal{M} \subseteq \mathbb{R}^n$ *is a* complementary observability subspace *if* $\mathcal{M} \in \mathbb{S}$ *and the spectrum of* $(A + JC)/\mathcal{M}$ *can be freely assigned by suitably choosing* $J \in \boldsymbol{J}(\mathcal{M})$. *The set of all complementary observability subspaces is denoted as* $\mathbb{M}$.

Any complementary observability subspace is a complementary detectability subspace (independent of the choice of $\mathbb{C}_g$), i.e., $\mathbb{M} \subseteq \mathbb{S}_g \subseteq \mathbb{S}$.

The sets $\mathbb{M}$, $\mathbb{S}_g$, and $\mathbb{S}$ are all closed under subspace intersection (see e.g. Willems and Commault, 1981, Section 4). As a consequence, given any set $\mathcal{Y} \subseteq \mathbb{R}^n$, there exist members $\mathcal{S}^*_{\mathcal{Y}}$, $\mathcal{S}^*_{g,\mathcal{Y}}$, and $\mathcal{M}^*_{\mathcal{Y}}$ of $\mathbb{S}$, $\mathbb{S}_g$, and $\mathbb{M}$ respectively, that are minimal subject to containing $\mathcal{Y}$. As a consequence of the relation $\mathbb{M} \subseteq \mathbb{S}_g \subseteq \mathbb{S}$, we have that $\mathcal{Y} \subseteq \mathcal{S}^*_{\mathcal{Y}} \subseteq \mathcal{S}^*_{g,\mathcal{Y}} \subseteq \mathcal{M}^*_{\mathcal{Y}} \subseteq \mathbb{R}^n$.

Conditionally invariant subspaces are closely related to the construction of observers. Assume $\mathcal{S} \in \mathbb{S}$. Consider the observer

$$\dot{z}(t) = [(A + JC)/\mathcal{S}]z(t) + [(Bu(t)) + \mathcal{S}] - [(Jy(t)) + \mathcal{S}] \tag{C.2}$$

with state vector $z(t) \in \mathbb{R}^n/\mathcal{S}$ for the system (C.1). Define the observer error $\varepsilon(t) = (x(t) + \mathcal{S}) - z(t)$. Then

$$\begin{aligned}\dot{\varepsilon}(t) &= [(Ax(t) + Bu(t)) + \mathcal{S}] - [(A + JC)/\mathcal{S}]z(t) - [(Bu(t)) + \mathcal{S}] + [(Jy(t)) + \mathcal{S}] \\ &= [(A + JC)/\mathcal{S}]\varepsilon(t),\end{aligned}$$

i.e., the observer error dynamics are governed by $(A + JC)/\mathcal{S}$. The spectrum of the observer error dynamics can be chosen to be contained in $\mathbb{C}_g$ if and only if $\mathcal{S} \subseteq \mathbb{S}_g$.

C.2.2 Unknown Input Observers

The methods proposed in Chapter 4 require unknown input observers, i.e., dynamic observers that yield asymptotic estimates despite the system inputs being unknown.

In the above construction, the observer dynamics (C.2) are independent of the input $u(t)$ if $\text{im}(B) \subseteq \mathcal{S}$. This suggests that, for some matrix $D \in \mathbb{R}^{r\times n}$, an unknown-input observer that yields an asymptotic estimate $\hat{w}(t)$ for the quantity $w(t) = Dx(t) \in \mathbb{R}^r$ with spectrum of the observer error dynamics contained in $\mathbb{C}_g$ exists if $\mathcal{S}_{g,\text{im}(B)} \subseteq \ker(D)$. Since trivial observers exist for $w(t) = Cx(t)$, it actually suffices if $\mathcal{S}_{g,\text{im}(B)} \bigcap \ker(C) \subseteq \ker(D)$. This condition is also necessary as stated in the following lemma due to Willems and Commault (1981, Proposition 4), which is repeated here without proof:

Lemma C.4. *Let system* (C.1) *be described by matrices $A \in \mathbb{R}^{n\times n}$, $B \in \mathbb{R}^{n\times p}$, $C \in \mathbb{R}^{q\times n}$. Let $D \in \mathbb{R}^{r\times n}$ and let $\mathbb{C} = \mathbb{C}_g \bigcup \mathbb{C}_b$, $\mathbb{C}_g \bigcap \mathbb{C}_b = \emptyset$ be a symmetric partitioning of the complex plane.*

There exists an unknown-input observer

$$\begin{aligned}\dot{z}(t) &= Ez(t) + Fy(t) \\ \hat{\zeta}(t) &= Rz(t) + Sy(t)\end{aligned}$$

with state vector $z(t) \in \mathbb{R}^m$ and estimate $\hat{\zeta}(t) \in \mathbb{R}^r$ for $\zeta(t) = Dx(t)$ such that the spectrum of the observer error dynamics is contained in $\mathbb{C}_g$ if and only if $\mathcal{S}^*_{g,\text{im}(B)} \bigcap \ker(C) \subseteq \ker(D)$.

In order to verify the conditions of Lemma C.4 or to determine which part of the state can be reconstructed with an unknown-input observer, we need to construct the subspace $\mathcal{S}^*_{g,\text{im}(B)}$. To this end, we first need the subspaces $\mathcal{S}^*_{\text{im}(B)}$ and $\mathcal{M}^*_{\text{im}(B)}$. Those subspaces can be obtained with the following algorithms that are proved e.g. in Willems and Commault (1981):

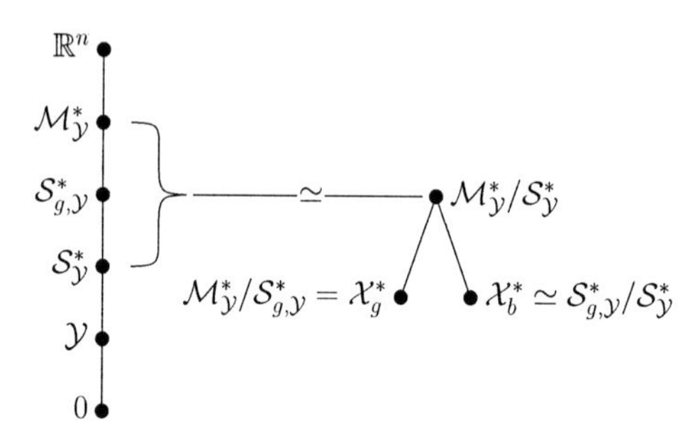

Figure C.2: Lattice diagrams for the set inclusion conditions involved in the construction of $\mathcal{S}^*_{g,\mathcal{Y}}$.

Algorithm C.5 (Willems and Commault 1981, Algorithm ISA'). *Consider the sequence* $\{\mathcal{S}^{\mu}_{\mathcal{Y}}\}$, $\mu \in \mathbb{N} \bigcup \{0\}$ *with*

$$\mathcal{S}^0_{\mathcal{Y}} = \{0\},$$
$$\mathcal{S}^{\mu}_{\mathcal{Y}} = \mathcal{Y} + A\left(\mathcal{S}^{\mu-1}_{\mathcal{Y}} \bigcap \ker(C)\right), \qquad \mu \in \mathbb{N}.$$

This sequence converges, strictly increasingly, in a finite number of steps and $\mathcal{S}^{\infty}_{\mathcal{Y}} = \mathcal{S}^*_{\mathcal{Y}}$ *is the member of* $\mathbb{S}$ *which is minimal subject to containing* $\mathcal{Y}$.

Algorithm C.6 (Willems and Commault 1981, Algorithm ACSA'). *Consider the sequence* $\{\mathcal{M}^{\mu}_{\mathcal{Y}}\}, \mu \in \mathbb{N} \bigcup \{0\}$ *with*

$$\mathcal{M}^0_{\mathcal{Y}} = \mathbb{R}^n,$$
$$\mathcal{M}^{\mu}_{\mathcal{Y}} = \mathcal{S}^*_{\mathcal{Y}} + (A^{-1}\mathcal{M}^{\mu-1}_{\mathcal{Y}}) \bigcap \ker(C), \qquad \mu \in \mathbb{N}.$$

This sequence converges, strictly decreasingly, in a finite number of steps and $\mathcal{M}^{\infty}_{\mathcal{Y}} = \mathcal{M}^*_{\mathcal{Y}}$ *is the member of* $\mathbb{M}$ *which is minimal subject to containing* $\mathcal{Y}$.

The procedure to obtain $\mathcal{S}^*_{g,\mathcal{Y}}$ is obtained from Wonham (1985, Section 5.6) using duality and can be summarized as follows (cf. Figure C.2): Choose any friend $J \in \boldsymbol{J}(\mathcal{S}^*_{\mathcal{Y}})$. By Wonham (1985, Corollary 5.1) and duality, we have $J \in \boldsymbol{J}(\mathcal{M}^*_{\mathcal{Y}})$, i.e., J is also a friend of $\mathcal{M}^*_{\mathcal{Y}}$. Let P be the canonical projection $P : \mathbb{R}^n \to \mathbb{R}^n/\mathcal{S}^*_{\mathcal{Y}}$. Define the maps $A^J \triangleq (A+JC)$ and $\overline{A}^J \triangleq A^J/\mathcal{S}^*_{\mathcal{Y}}$. The subspace $\mathcal{M}^*_{\mathcal{Y}}/\mathcal{S}^*_{\mathcal{Y}} \subseteq \mathbb{R}^n/\mathcal{S}^*_{\mathcal{Y}}$ is $\overline{A}^J$ invariant for

$$\overline{A}^J(\mathcal{M}^*_{\mathcal{Y}}/\mathcal{S}^*_{\mathcal{Y}}) = \overline{A}^J P\mathcal{M}^*_{\mathcal{Y}} = PA^J\mathcal{M}^*_{\mathcal{Y}} \subseteq P\mathcal{M}^*_{\mathcal{Y}} = \mathcal{M}^*_{\mathcal{Y}}/\mathcal{S}^*_{\mathcal{Y}}.$$

Furthermore, using a duality argument, it follows from Wonham (1985, Theorem 5.7) that the restriction $\overline{A}^J|\mathcal{M}^*_{\mathcal{Y}}/\mathcal{S}^*_{\mathcal{Y}} : \mathcal{M}^*_{\mathcal{Y}}/\mathcal{S}^*_{\mathcal{Y}} \to \mathcal{M}^*_{\mathcal{Y}}/\mathcal{S}^*_{\mathcal{Y}}$ of $\overline{A}^J$ to $\mathcal{M}^*_{\mathcal{Y}}/\mathcal{S}^*_{\mathcal{Y}}$ is independent of $J \in \boldsymbol{J}(\mathcal{S}^*_{\mathcal{Y}})$. Thus, there exist subspaces $\mathcal{X}^*_b \subseteq \mathcal{M}^*_{\mathcal{Y}}/\mathcal{S}^*_{\mathcal{Y}}$ and $\mathcal{X}^*_g \subseteq \mathcal{M}^*_{\mathcal{Y}}/\mathcal{S}^*_{\mathcal{Y}}$ independent of $J \in \boldsymbol{J}(\mathcal{S}_{\mathcal{Y}}^*)$ with the property that $\overline{A}^J \mathcal{X}_b^* \subseteq \mathcal{X}_b^*$ and $\overline{A}^J \mathcal{X}_g^* \subseteq \mathcal{X}_g^*$, $\sigma\left(\overline{A}^J | \mathcal{X}_b^*\right) \subseteq \mathbb{C}_b$ and $\sigma\left(\overline{A}^J | \mathcal{X}_g^*\right) \subseteq \mathbb{C}_g$, and $\mathcal{X}_b^* \oplus \mathcal{X}_g^* = \mathcal{M}_{\mathcal{Y}}^* / \mathcal{S}_{\mathcal{Y}}^*$. We have

$$\mathcal{S}_{g,\mathcal{Y}}^* = P^{-1} \mathcal{X}_b^*. \tag{C.3}$$

It is sometimes more convenient to decompose the entire state space $\mathbb{R}^n$ instead of the space $\mathcal{M}_{\mathcal{Y}}^* / \mathcal{S}_{\mathcal{Y}}^*$. There always exist subspaces $\mathcal{X}_b \subseteq \mathbb{R}^n$ and $\mathcal{X}_g \subseteq \mathbb{R}^n$ with the property that $A^J \mathcal{X}_b \subseteq \mathcal{X}_b$ and $A^J \mathcal{X}_g \subseteq \mathcal{X}_g$, $\sigma(A^J | \mathcal{X}_b) \subseteq \mathbb{C}_b$ and $\sigma(A^J | \mathcal{X}_g) \subseteq \mathbb{C}_g$, and $\mathcal{X}_b \oplus \mathcal{X}_g = \mathbb{R}^n$. These subspaces generally depend on $J \in \boldsymbol{J}(\mathcal{S}_{\mathcal{Y}}^*)$. Yet, they are related to the subspaces $\mathcal{X}_b^* \subseteq \mathcal{M}_{\mathcal{Y}}^* / \mathcal{S}_{\mathcal{Y}}^*$ and $\mathcal{X}_g^* \subseteq \mathcal{M}_{\mathcal{Y}}^* / \mathcal{S}_{\mathcal{Y}}^*$ as $P(\mathcal{X}_b \bigcap \mathcal{M}_{\mathcal{Y}}^*) = \mathcal{X}_b^*$ and $P(\mathcal{X}_g \bigcap \mathcal{M}_{\mathcal{Y}}^*) = \mathcal{X}_g^*$. It is thus possible to replace (C.3) and determine $\mathcal{S}_{g,\mathcal{Y}}^*$ as

$$S_{g,\mathcal{Y}}^* = \mathcal{S}_{\mathcal{Y}}^* + \mathcal{X}_b \bigcap \mathcal{M}_{\mathcal{Y}}^*. \tag{C.4}$$

The latter equation has the advantage that it can be used to determine $\mathcal{S}_{g,\mathcal{Y}}^*$ without computing the maps $\overline{A}^J$ and $\overline{A}^J | \mathcal{M}_{\mathcal{Y}}^* / \mathcal{S}_{\mathcal{Y}}^*$, which are needed to determine the set $\mathcal{X}_b^*$ involved in (C.3).

Bibliography

J. A. Acebrón, L. L. Bonilla, C. J. P. Vicente, F. Ritort, and R. Spigler. The Kuramoto model: A simple paradigm for synchronization phenomena. *Reviews of Modern Physics*, 77(1):137–185, 2005.

B. W. Andrews, P. A. Iglesias, and E. D. Sontag. Signal detection and approximate adaptation implies an approximate internal model. In *Proceedings of the 45th IEEE Conference on Decision and Control*, pages 2364–2369, 2006.

B. W. Andrews, E. D. Sontag, and P. A. Iglesias. An approximate internal model principle: Applications to nonlinear models of biological systems. In *Proceedings of the 17th IFAC World Congress*, page FrB25.3, 2008.

T. M. Apostol. *Mathematical Analysis.* Addison-Wesley, 2nd edition, 1974.

M. Arcak. Passivity as a design tool for group coordination. *IEEE Transactions on Automatic Control*, 52(8):1380–1390, 2007.

H. Bai, M. Arcak, and J. T. Wen. Adaptive motion coordination: Using relative velocity feedback to track a reference velocity. *Automatica*, 45(4):1020–1025, 2009.

R. S. Baron, N. L. Kerr, and N. Miller. *Group Process, Group Decision, Group Action.* Mapping Social Psychology. Open University Press, 1992.

M. Bennett, M. F. Schatz, H. Rockwood, and K. Wiesenfeld. Huygens's clocks. *Proceedings of the Royal Society A*, 458(2019):563–579, 2002.

I. I. Blekhman, A. L. Fradkov, H. Nijmeijer, and A. Y. Pogromsky. On self-synchronization and controlled synchronization. *Systems and Control Letters*, 31:299–305, 1997.

B. Bollobás. *Modern Graph Theory*, volume 184 of *Graduate Texts in Mathematics.* Springer, 1998.

S. P. Boyd and L. Vandenberghe. *Convex Optimization.* Cambridge University Press, 2004.

S. P. Boyd, L. El Ghaoui, E. Feron, and V. Balakrishnan. *Linear Matrix Inequalities in System and Control Theory*, volume 15 of *SIAM Studies in Applied Mathematics.* SIAM, 1994.

S. P. Boyd, A. Ghosh, B. Prabhakar, and D. Shah. Gossip algorithms: Design, analysis and applications. In *Proceedings of the 24th IEEE Conference on Computer Communications*, pages 1653–1664, 2005.

M. Bürger and M. Guay. Robust constraint satisfaction for continuous-time nonlinear systems. In *Proceedings of the 47th IEEE Conference on Decision and Control*, pages 7–12, 2008.

J. B. Buck. Synchronous rhythmic flashing of fireflies. *The Quaterly Review of Biology*, 13(3):301–314, 1938.

J. B. Buck. Synchronous rhythmic flashing of fireflies. II. *The Quaterly Review of Biology*, 63(3):265–289, 1988.

A. R. Bulsara, T. C. Elston, C. R. Doering, S. B. Lowen, and K. Lindenberg. Cooperative behavior in periodically driven noisy integrate-fire models of neuronal dynamics. *Physical Review E*, 53(4):3958–3969, 1996.

C. I. Byrnes and A. Isidori. Output regulation for nonlinear systems: An overview. In *Proceedings of the 37th IEEE Conference on Decision and Control*, pages 3069–3074, 1998.

C. I. Byrnes and A. Isidori. Limit sets, zero dynamics, and internal models in the problem of nonlinear output regulation. *IEEE Transactions on Automatic Control*, 48(10):1712–1723, 2003.

C. I. Byrnes, A. Isidori, and J. C. Willems. Passivity, feedback equivalence, and the global stabilization of minimum phase nonlinear systems. *IEEE Transactions on Automatic Control*, 36(11):1228–1240, 1991.

C. I. Byrnes, F. Delli Priscoli, and A. Isidori. *Output Regulation of Uncertain Nonlinear Systems*. Systems and Control: Foundations and Applications. Birkhäuser, 1997.

C. C. Canavier and S. Achuthan. Pulse coupled oscillators. *Scholarpedia*, 2(4):1331, 2007. http://www.scholarpedia.org/article/Pulse_coupled_oscillators.

Y.-Y. Cao, J. Lam, and Y.-X. Sun. Static output feedback stabilization: An ILMI approach. *Automatica*, 34(12):1641–1645, 1998.

R. Carli, F. Fagnani, A. Speranzon, and S. Zampieri. Communication constraints in the average consensus problem. *Automatica*, 44(3):671–684, 2008.

D. E. Chang, S. C. Shadden, J. E. Marsden, and R. Olfati-Saber. Collision avoidance for multiple agent systems. In *Proceedings of the 42nd IEEE Conference on Decision and Control*, pages 539–543, 2003.

N. Chopra and M. W. Spong. Passivity-based control of multi-agent systems. In S. Kawamura and M. Svinin, editors, *Advances in Robot Control, From Everyday Physics to Human-Like Movements*, pages 107–134. Springer, 2006.

E. A. Coddington and N. Levinson. *Theory of Ordinary Differential Equations*. International Series in Pure and Applied Mathematics. McGraw-Hill Book Company, 1955.

L. von Collatz and U. Sinogowitz. Spektren endlicher Grafen. *Abhandlungen aus dem Mathematischen Seminar der Universität Hamburg*, 21(1):63–77, 1957. In German.

J. Cortés. Distributed algorithms for reaching consensus on general functions. *Automatica*, 44(3):726–737, 2008.

M. Darouach, M. Zasadzinski, and S. J. Xu. Full-order observers for linear systems with unknown inputs. *IEEE Transactions on Automatic Control*, 39(3):606–609, 1994.

R. Diestel. *Graph Theory*, volume 173 of *Graduate Texts in Mathematics*. Springer, 3rd edition, 2005.

J. A. Fax and R. M. Murray. Graph Laplacians and stabilization of vehicle formations. In *Proceedings of the 15th IFAC World Congress*, 2002.

J. A. Fax and R. M. Murray. Information flow and cooperative control of vehicle formations. *IEEE Transactions on Automatic Control*, 49(9):1465–1476, 2004.

N. Fenichel. Persistence and smoothness of invariant manifolds for flows. *Indiana University Mathematics Journal*, 21:193–226, 1972.

M. Fiedler. Algebraic connectivity of graphs. *Czechoslovak Mathematical Journal*, 23(2): 298–305, 1973.

M. Fiedler. Property of eigenvectors of nonnegative symmetric matrices and its application to graph theory. *Czechoslovak Mathematical Journal*, 25(4):619–633, 1975.

B. A. Francis. The linear multivariable regulator problem. *SIAM Journal on Control and Optimization*, 15(3):486–505, 1977.

B. A. Francis and W. M. Wonham. The internal model principle for linear multivariable regulators. *Applied Mathematics and Optimization*, 2(2):170–194, 1975.

B. A. Francis and W. M. Wonham. The internal model principle of control theory. *Automatica*, 12:457–465, 1976.

M. P. J. Fromherz and W. B. Jackson. Force allocation in a large-scale distributed active surface. *IEEE Transactions on Control Systems Technology*, 11(5):641–655, 2003.

M. C. de Gennaro and A. Jadbabaie. Decentralized control of connectivity for multi-agent systems. In *Proceedings of the 45th IEEE Conference on Decision and Control*, pages 3628–3633, 2006.

L. E. Ghaoui, F. Oustry, and M. AitRami. A cone complementarity linearization algorithm for static output-feedback and related problems. *IEEE Transactions on Automatic Control*, 42(8):1171–1176, 1997.

C. Godsil and G. Royle. *Algebraic Graph Theory*, volume 207 of *Graduate Texts in Mathematics*. Springer, 2004.

R. Goebel, R. G. Sanfelice, and A. R. Teel. Hybrid dynamical systems. *IEEE Control Systems Magazine*, 29(2):28–93, 2009.

C. V. Goldman and S. Zilberstein. Optimizing information exchange in cooperative multiagent systems. In *Proceedings of the 2nd international Conference on Autonomous Agents and Multiagent Systems*, 2003.

J. Grasman, F. Verhulst, and S.-D. Shih. The Lyapunov exponents of the Van der Pol oscillator. *Mathematical Methods in the Applied Science*, 28(10):1131–1139, 2005.

W. Hahn. *Stability of Motion.* Springer, 1967.

J. K. Hale. *Asymptotic Behavior of Disspative Systems*, volume 25 of *Mathematical Surveys and Monographs.* American Mathematical Society, 1988.

J. K. Hale. Diffusive coupling, dissipation, and synchronization. *Journal of Dynamics and Differential Equations*, 9(1):1–52, 1997.

J. K. Hale, L. T. Magalhães, and W. M. Oliva. *Dynamics in Infinite Dimensions*, volume 47 of *Applied Mathematical Sciences.* Springer, 2nd edition, 2002.

P. Hartman. *Ordinary Differential Equations.* John Wiley and Sons, 1964.

M. Hautus. Linear matrix equations with applications to the regulator problem. In I. Landau, editor, *Outils et modèles mathématiques pour l'automatique, l'analyse de système et le traitement du signal*, volume 3, pages 399–412. C.N.R.S. Paris, 1983.

D. Helbing, I. Farkas, and T. Vicsek. Simulating dynamical features of escape panic. *Nature*, 407:487–490, 2000.

J. P. Hespanha, P. Naghshtabrizi, and Y. Xu. A survey of recent results in networked control systems. *Proceedings of the IEEE*, 95(1):138–162, 2007.

M. D. Hill. What is scalability? *SIGARCH Computer Architecture News*, 18(4):18–21, 1990.

F. C. Hoppensteadt and E. M. Izhikevich. *Weakly Connected Neural Networks*, volume 126 of *Applied Mathematical Sciences.* Springer, 1997.

R. A. Horn and C. R. Johnson. *Matrix Analysis.* Cambridge University Press, 1985.

R. A. Horn and C. R. Johnson. *Topics in Matrix Analysis.* Cambridge University Press, 1991.

J. Hu and Y. S. Lin. Consensus control for multi-agent systems with double-integrator dynamics and time delays. *IET Control Theory and Applications*, 4(1):109–118, 2010.

F. Hutu, S. Cauet, and P. Coirault. Robust synchronization of different coupled oscillators: Application to antenna arrays. *Journal of the Franklin Institute*, 346(5):413–430, 2009.

A. Isidori. *Nonlinear Control Systems.* Springer, 3rd edition, 1995.

A. Isidori and C. I. Byrnes. Steady-state behaviors in nonlinear systems with an application to robust disturbance rejection. *Annual Reviews in Control*, 32:1–16, 2008.

E. M. Izhikevich. *Dynamical Systems in Neuroscience: The Geometry of Excitability and Bursting.* Computational Neuroscience. MIT Press, 2007.

E. M. Izhikevich and G. B. Ermentrout. Phase model. *Scholarpedia*, 3(10):1487, 2008. http://www.scholarpedia.org/article/Phase_model.

A. Jadbabaie, J. Lin, and A. S. Morse. Coordination of groups of mobile autonomous agents using nearest neighbor rules. In *Proceedings of the 41st IEEE Conference on Decision and Control*, pages 2953–2958, 2002.

A. Jadbabaie, J. Lin, and A. S. Morse. Coordination of groups of mobile autonomous agents using nearest neighbor rules. *IEEE Transactions on Automatic Control*, 48(6): 988–1001, 2003.

T. Kailath. *Linear Systems*. Prentice Hall, 1980.

H. K. Khalil. *Nonlinear Systems*. Prentice Hall, 3rd edition, 2002.

H. Kim, H. Shim, and J. H. Seo. Output consensus of heterogeneous uncertain linear multi-agent systems. IEEE Transactions on Automatic Control, 2010. To appear.

Y. Kim and M. Mesbahi. On maximizing the second smallest eigenvalue of a state-dependent graph Laplacian. *IEEE Transactions on Automatic Control*, 51(1):116–120, 2006.

H. W. Knobloch and B. Aulbach. The role of center manifolds in ordinary differential equations. In *Proceedings of the 5th Chechoslovak Conference of Differential Equations and Applications*, pages 179–189. Teubner, 1981.

H. W. Knobloch, A. Isidori, and D. Flockerzi. *Topics in Control Theory*. Birkhäuser, 1993.

N. Kopell and G. B. Ermentrout. Symmetry and phaselocking in chains of weakly coupled oscillators. *Communications on Pure and Applied Mathematics*, 39(5):623–660, 1986.

A. J. Krener. Feedback linearization. In J. Baillieul and J. C. Willems, editors, *Mathematical Control Theory*, pages 66–98. Springer, 1999.

Y. Kuramoto. Self-entrainment of a population of coupled non-linear oscillators. In H. Araki, editor, *International Symposium on Mathematical Problems in Theoretical Physics*, volume 39 of *Lecture Notes in Physics*, pages 420–422. Springer, 1975.

S. Lang. *Calculus of Several Variables*. Undergraduate Texts in Mathematics. Springer, 1987.

D. Lee and M. W. Spong. Stable flocking of multiple inertial agents on balanced graphs. *IEEE Transactions on Automatic Control*, 52(8):1469–1475, 2007.

F. Leibfritz. An LMI-based algorithm for designing suboptimal static $\mathcal{H}_2/\mathcal{H}_\infty$ output feedback controllers. *SIAM Journal on Control and Optimization*, 39(6):1711–1735, 2001.

Z. Lin. *Coupled Dynamical Systems: From Structure Towards Stability and Stabilization*. PhD thesis, University of Toronto, Graduate Department of Electrical and Computer Engineering, 2006.

A.-M. Liénard. Etude des oscillations entretenues. *Revue générale de l'électricité*, 23: 901–912, 946–954, 1928.

O. L. Mangasarian and J.-S. Pang. The extended linear complementarity problem. *SIAM Journal on Matrix Analysis and Applications*, 16(2):359–368, 1995.

A. Mauroy and R. Sepulchre. Clustering behaviors in networks of integrate-and-fire oscillators. *Chaos*, 18:037122, 2008.

R.-M. Memmesheimer and M. Timme. Designing complex networks. *Physica D*, 224(1-2): 182–201, 2006.

R. Merris. A survey of graph Laplacians. *Linear and Multilinear Algebra*, 39:19–31, 1995.

R. E. Mirollo and S. H. Strogatz. Synchronization of pulse-coupled biological oscillators. *SIAM Journal on Applied Mathematics*, 50(6):1645–1662, 1990.

K. Mischaikow, H. Smith, and H. R. Thieme. Asymptotically autonomous semiflows: Chain reccurence and Lyapunov functions. *Transactions of the American Mathematical Society*, 347(5):1669–1685, 1995.

B. Mohar. Laplace eigenvalues of graphs – a survey. *Discrete Mathematics*, 109(1-3): 171–183, 1992.

L. Moreau. Stability of continuous-time distributed consensus algorithms. http://arxiv.org/abs/math/0409010v1, 2004a. arXiv:math/0409010v1 [math.OC].

L. Moreau. Stability of continuous-time distributed consensus algorithms. In *Proceedings of the 43rd IEEE Conference on Decision and Control*, pages 3998–4003, 2004b.

L. Moreau. Stability of multiagent systems with time-dependent communication links. *IEEE Transactions on Automatic Control*, 50(2):169–182, 2005.

S. Nair and N. E. Leonard. Stable synchronization of mechanical system networks. *SIAM Journal on Control and Optimization*, 47(2):234–265, 2008.

M. W. Newman. *The Laplacian Spectrum of Graphs*. PhD thesis, Department of Mathematics, University of Manitoba, Winnipeg, Canada, 2000.

H. Nijmeijer and I. M. Y. Mareels. An observer looks at synchronization. *IEEE Transactions on Circuits and Systems—Part I: Fundamental Theory and Applications*, 44(10): 882–890, 1997.

R. Olfati-Saber. Flocking for multi-agent dynamic systems: Algorithms and theory. *IEEE Transactions on Automatic Control*, 51(3):401–420, 2006.

R. Olfati-Saber and R. M. Murray. Consensus problems in networks of agents with switching topology and time-delays. *IEEE Transactions on Automatic Control*, 49(9):1520–1533, 2004.

R. Olfati-Saber, J. A. Fax, and R. M. Murray. Consensus and cooperation in networked multi-agent systems. *Proceedings of the IEEE*, 95(1):215–233, 2007.

A. Papachristodoulou, A. Jadbabaie, and U. Münz. Effects of delay in multi-agent consensus and oscillator synchronization. *IEEE Transactions on Automatic Control*, 55(6): 1471–1477, 2010.

A. Pavlov, N. van de Wouw, and H. Nijmeijer. *Uniform Output Regulation of Nonlinear Systems, A Convergent Dynamics Approach.* Systems and Control: Foundations and Applications. Birkhäuser, 2006.

C. S. Peskin. *Mathematical Aspects of Heart Physiology*, chapter VII.7 Self Synchronization of the Cardiac Pacemaker, pages 268–278. Courant Institute of Mathematical Sciences, New York University, 1975.

A. Pikovsky, M. Rosenblum, and J. Kurths. *Synchronization. A universal concept in nonlinear sciences*, volume 12 of *Cambridge Nonlinear Science Series.* Cambridge University Press, 2001.

A. Y. Pogromsky. Passivity based design of synchronizing systems. *International Journal of Bifurcation and Chaos*, 8(2):295–319, 1997.

Z. Qu, J. Wang, and R. A. Hull. Cooperative control of dynamical systems with application to autonomous vehicles. *IEEE Transactions on Automatic Control*, 53(4):894–911, 2008.

W. Ren. Multi-vehicle consensus with a time-varying reference state. *Systems and Control Letters*, 56(7-8):474–483, 2007.

W. Ren. Synchronization of coupled harmonic oscillators with local interaction. *Automatica*, 44(12):3195–3200, 2008.

W. Ren and E. Atkins. Second-order consensus protocols in multiple vehicle systems with local interactions. In *Proceedings of the AIAA Guidance, Navigation, and Control Conference*, pages AIAA–2005–6238, 2005.

W. Ren and R. W. Beard. Consensus seeking in multiagent systems under dynamically changing interaction topology. *IEEE Transactions on Automatic Control*, 50(5):655–661, 2005.

W. Ren, R. W. Beard, and T. W. McLain. Coordination variables and consensus building in multiple vehicle systems. In V. Kumar, N. Leonard, and A. S. Morse, editors, *Cooperative Control*, volume 309 of *Lecture Notes in Control and Information Sciences*, pages 171–188. Springer, 2004.

W. Ren, K. Moore, and Y. Chen. High-order consensus algorithms in cooperative vehicle systems. In *Proceedings of the 2006 IEEE International Conference on Networking, Sensing and Control*, pages 457– 462, 2006.

W. Ren, R. W. Beard, and E. M. Atkins. Information consensus in multivehicle cooperative control. *IEEE Control Systems Magazine*, 27(2):71–82, 2007.

C. W. Reynolds. Flocks, herds, and schools: A distributed behavioral model. In *Proceedings of the 14th Annual Conference on Computer Graphics and Interactive Techniques*, pages 25–34. ACM, 1987.

A. Sarlette. *Geometry and Symmetries in Coordination Control.* PhD thesis, Department of Electrical Engineering and Computer Science, University of Liège, Belgium, 2009.

A. Sarlette and R. Sepulchre. Synchronization on the circle. http://arxiv.org/abs/0901.2408v1, 2009a. arXiv:0901.2408v1 [math.OC].

A. Sarlette and R. Sepulchre. Consensus on homogeneous manifolds. In *Proceedings of the joint 48th IEEE Conference on Decision and Control and Chinese Control Conference*, pages 6438–6443, 2009b.

A. Sarlette, C. Bastin, M. Dimmler, B. Sedghi, T. Erm, B. Bauvir, and R. Sepulchre. Integral control from distributed sensing: an extremely large telescope case study. *IEEE Transactions on Control Systems Technology*, 2010. Submitted.

L. Scardovi and R. Sepulchre. Synchronization in networks of identical linear systems. http://arxiv.org/abs/0805.3456, 2008. arXiv:0805.3456v1 [math.OC].

L. Scardovi and R. Sepulchre. Synchronization in networks of identical linear systems. *Automatica*, 44(11):2557–2562, 2009.

L. Scardovi, M. Arcak, and E. D. Sontag. Synchronization of interconnected systems with applications to biochemical networks: an input-output approach. *IEEE Transactions on Automatic Control*, 55(6):1367–1379, 2010.

J. H. Seo, H. Shim, and J. Back. Consensus of high-order linear systems using dynamic output feedback compensator: Low gain approach. *Automatica*, 45(11):2659–2664, 2009a.

J. H. Seo, H. Shim, and J. Back. Consensus and synchronization of linear high-order systems via output coupling. In *Proceedings of the European Control Conference*, pages 767–772, 2009b.

R. Sepulchre, M. Jankovic, and P. Kokotovic. *Constructive Nonlinear Control*. Springer, 1997.

R. Sepulchre, D. Paley, and N. E. Leonard. Collective motion and oscillator synchronization. In V. Kumar, N. Leonard, and A. Morse, editors, *Cooperative control*, volume 309 of *Lecture Notes in Control and Information Sciences*, pages 189–205. Springer, 2004.

R. Sepulchre, D. A. Paley, and N. E. Leonard. Stabilization of planar collective motion with limited communication. *IEEE Transactions on Automatic Control*, 53(3):706–719, 2008.

T. J. Slight, B. Romeira, L. Wang, J. M. L. Figueiredo, E. Wasige, and C. N. Ironside. A Liénard oscillator resonant tunnelling diode-laser diode hybrid integrated circuit: Model and experiment. *IEEE Journal of Quantum Electronics*, 44(12):1158–1163, 2008.

J.-J. E. Slotine and W. Wang. A study of synchronization and group cooperation using partial contraction theory. In *Cooperative Control*, volume 309 of *Lecture Notes in Control and Information Sciences*, pages 207–228. Springer, 2004.

G.-B. Stan and R. Sepulchre. Analysis of interconnected oscillators by dissipativity theory. *IEEE Transactions on Automatic Control*, 52(2):256–270, 2007.

A. Strauss and J. A. Yorke. On asymptotically autonomous differential equations. *Mathematical Systems Theory*, 1(2):175–182, 1967.

S. Strogatz. *Sync. The emerging science of spontaneous order.* Penguin Books, 2003.

Y. G. Sun, L. Wang, and G. Xie. Average consensus in networks of dynamic agents with switching topologies and multiple time-varying delays. *Systems and Control Letters*, 57 (2):175–183, 2008.

A. Tahbaz-Salehi and A. Jadbabaie. A necessary and sufficient condition for consensus over random networks. *IEEE Transactions on Automatic Control*, 53(3):791–795, 2008.

H. G. Tanner, A. Jadbabaie, and G. J. Pappas. Coordination of multiple autonomous vehicles. In *Proceedings of the 11th IEEE Mediterranean Conference on Control and Automation*, 2003.

H. G. Tanner, G. J. Pappas, and V. Kumar. Leader-to-formation stability. *IEEE Transactions on Robotics and Automation*, 20(3):443–455, 2004.

H. G. Tanner, A. Jadbabaie, and G. J. Pappas. Flocking in fixed and switching networks. *IEEE Transactions on Automatic Control*, 52(5):863–868, 2007.

H. L. Trentelman, A. A. Stoorvogel, and M. Hautus. *Control theory for linear systems.* Springer, 2001.

J. N. Tsitsiklis. *Problems in Decentralized Decision Making and Computation.* PhD thesis, Department of Electrical Engineering and Computer Science, Massachusetts Institute of Technology, 1984.

S. E. Tuna. LQR-based coupling gain for synchronization of linear systems. http://arxiv.org/abs/0801.3390, 2008a. arXiv:0801.3390v1 [math.OC].

S. E. Tuna. Conditions for synchronizability in arrays of coupled linear systems. http://arxiv.org/abs/0811.3530, 2008b. arXiv:0811.3530v1 [math.DS].

T. Vicsek, A. Czirók, E. Ben-Jacob, I. Cohen, and O. Shochet. Novel type of phase transition in a system of self-driven particles. *Physical Review Letters*, 75(7):1226–1229, 1995.

M. Vidyasagar. *Input-Output Analysis of Large-Scale Interconnected Systems. Decomposition, Well-Posedness and Stability*, volume 29 of *Lecture Notes in Control and Information Sciences.* Springer, 1981.

J. Wang, D. Cheng, and X. Hu. Consensus of multi-agent linear dynamic systems. *Asian Journal of Control*, 10(2):144–155, 2008.

W. Wang and J.-J. E. Slotine. On partial contraction analysis for coupled nonlinear oscillators. *Biological Cybernetics*, 92(1):38–53, 2005.

P. Wieland and F. Allgöwer. Constructive safety using control barrier functions. In *Proceedings of the 7th IFAC Symposium on Nonlinear Control Systems*, pages 473–478, 2007.

P. Wieland and F. Allgöwer. An internal model principle for consensus in heterogeneous linear multi-agent systems. In *Proceedings of the 1st IFAC Workshop on Distributed Estimation and Control in Networked Systems*, pages 7–12, 2009a.

P. Wieland and F. Allgöwer. An internal model principle for synchronization. In *Proceedings of the 7th IEEE International Conference on Control and Automation*, pages 285–290, 2009b.

P. Wieland and F. Allgöwer. On consensus among identical linear systems using input-decoupled functional observers. In *Proceedings of the 2010 American Control Conference*, pages 1641–1646, 2010.

P. Wieland, C. Ebenbauer, and F. Allgöwer. Ensuring task-independent safety for multi-agent systems by feedback. In *Proceedings of the 2007 American Control Conference*, pages 3880–3885, 2007.

P. Wieland, J.-S. Kim, H. Scheu, and F. Allgöwer. On consensus in multi-agent systems with linear high-order agents. In *Proceedings of the 17th IFAC World Congress*, pages 1541–1546, 2008.

P. Wieland, J.-S. Kim, and F. Allgöwer. On topology and dynamics of consensus among linear high-order agents. *International Journal of Systems Science*, 2010a. In print (available online, doi: 10.1080/00207721003658202).

P. Wieland, G. S. Schmidt, R. Sepulchre, and F. Allgöwer. Phase synchronization through entrainment by a consensus input. In *Proceedings of the 49th IEEE Conference on Decision and Control*, 2010b. Accepted for publication.

P. Wieland, R. Sepulchre, and F. Allgöwer. An internal model principle is necessary and sufficient for linear output synchronization. *Automatica*, 2010c. Provisionally accepted.

J. C. Willems and C. Commault. Disturbance decoupling by measurement feedback with stability or pole placement. *SIAM Journal on Control and Optimization*, 19(4):490–504, 1981.

W. M. Wonham. *Linear Multivariable Control - A Geometric Approach*, volume 10 of *Applications of Mathematics*. Springer, 3rd edition, 1985.

C. W. Wu. *Synchronization in Coupled Chaotic Circuits and Systems*, volume 41 of *World Scientific Series on Nonlinear Science*. World Scientific, 2002.

C. W. Wu. Algebraic connectivity of directed graphs. *Linear and Multilinear Algebra*, 53 (3):203–223, 2005.

B. Yang and H. Fang. Forced consensus in networks of double integrator systems with delayed input. *Automatica*, 46(3):629–632, 2010.

S. Zampieri. Trends in networked control systems. In *Proceedings of the 17th IFAC World Congress*, pages 2886–2894, 2008.

D. Z. Zhao, C. W. Li, and J. Ren. Speed synchronization of multiple induction motors with adjacent cross-coupling control. *IET Control Theory and Applications*, 4(1):119–128, 2010.